DATE DUE

OCT 27			

Discovery Books are published by:
Red Deer Press
813 MLT, 2500 University Dr. N.W.
Calgary, Alberta, Canada
T2N 1N4
www.reddeerpress.com

Designed by Royal Tyrrell Museum of Palaeontology.
Printed and bound in China by Stone Sapphire for Red Deer Press.

Cover Images
(large image): Brian Kooyman
(smaller images): Royal Tyrrell Museum; (except *Archaeopteryx*) Dennis Braman, Royal Tyrrell Museum; (and mammoth) Chad Shier

The publishers gratefully acknowledge the financial assistance of the Alberta Foundation for the Arts, the Canada Council, the Department of Communications and the University of Calgary.

National Library of Canada Cataloguing in Publication Data

Reading the Rocks: A Biography of Ancient Alberta/the Royal Tyrrell Museum of Palaeontology.

Includes index.

Previous edition title: *The Land Before Us: The Making Of Ancient Alberta*

ISBN 0-88995-283-3 (pbk.) – ISBN 0-88995-288-4 (bound)

1. Geology-Alberta. 2. Palaeontology-Alberta.

QE748.A4R42 2003   557.123   C2003-910709-4

TABLE OF CONTENTS

FOREWARD

Unravelling the creation of our province is an exciting and never-ending scientific study. Alberta has been drowned by salty seas, baked under subtropical suns and crushed beneath tons of ice. Life has endured through all these changes—constantly evolving. The Royal Tyrrell Museum's scientific mandate is to investigate and explain the history of life on this planet, especially as it relates to Alberta. Palaeontology is not just about dead things—it is a dynamic changing science that brings the past to life.

Reading the Rocks: A Biography of Ancient Alberta presents the prehistoric development of Alberta and the western interior of North America. We hope to communicate the excitement of the past to Albertans today, showing how the resources and life we take for granted came to be. The Tyrrell's scientists can be proud of their role in the discovery of our province's past.

Dr. Bruce G. Naylor
Director, Royal Tyrrell Museum

ACKNOWLEDGEMENTS

This book tells the geological story of Alberta, with reference to research by Royal Tyrrell Museum of Palaeontology scientists and research partners. Consulting scientists were unstinting in generosity and encouragement for the project. They provided information, and examined all copy and artwork to ensure the currency and accuracy of content. Any errors that survived their rigorous review are mine.

Many thanks to: Dennis Braman, Don Brinkman, Philip Currie, David Eberth, James Gardner, Paul Johnston, Bruce Naylor, Andy Neuman and Betsy Nicholls—all Tyrrell researchers—and to: Brian Chatterton (University of Alberta), Christopher Collom (Mount Royal College), Pat Druckenmiller (University of Calgary), Richard Fox (University of Alberta), Leonard Hills (University of Calgary), Richard McCrea (University of Alberta), Paul McNeil (University of Calgary), Mark Wilson (University of Alberta), and John-Paul Zonneveld (Geological Survey of Canada). Kevin Aulenback, Clive Coy, Kevin Kruger, Ken Kucher, Vien Lam, Jim McCabe, Tim Schowalter, Wendy Sloboda, and Darren Tanke—Tyrrell technical staff past and present who shared information about specimens they had collected and prepared—were also essential to the project.

The Tyrrell's Don Brinkman and Bruce Naylor deserve additional thanks, as do Kathryn Valentine and Lindsay Cook for keeping the project on track. Lindsay, assisted by Wendy Taylor, had the additional burden of chasing down images. Thanks also to Red Deer Press's Dennis Johnson.

On a personal note—thank you, Scott.

Contributors of photographs are credited alongside the corresponding image.

Specimen illustrations were provided by Donna Sloan. Maps, globes and diagrams were created by Luke Webster.

Graphic layout was completed by Maureen Johnston.

As more and more information about the ancient past becomes known, the geologic time scale is updated. Differences in opinion remain among scientists. Dates and geologic time periods referred to in this book are roughly based upon the Geological Society of America's 1999 Geologic Time Scale.

chapter one

PAST TIMES, PRESENT PLACES: ALBERTA'S STORY

Outside the Royal Tyrrell Museum of Palaeontology, weather and water unmake the landscape. Seventy-million-year-old rocks dissolve, millimetre by millimetre. The sand, dust, and gravel wash into streams, ponds, and the river to collect on older sediment layers where the river flows into Hudson Bay, far to the northeast. Today, as happened millions of years ago, new rock forms from ruins of older rock. The progression of rock layers up the walls of the Red Deer River valley recounts the long story of the life, death, and rebirth of the local landscape through time.

Millions of Years Ago	EON	ERA	PERIOD	EPOCH
			Quaternary	Holocene
				Pleistocene
1.8				Pliocene
		CENOZOIC	Tertiary	Miocene
				Oligocene
				Eocene
				Palaeocene
65			Cretaceous	
144	PHANEROZOIC	MESOZOIC	Jurassic	
200			Triassic	
251			Permian	
300			Carboniferous	
355		PALAEOZOIC	Devonian	
418			Silurian	
441			Ordovician	
490			Cambrian	
545				
	PRECAMBRIAN		Proterozoic	
2,500			Archean	
3,900			Hadean	

Stories about Alberta date back to more than 12,000 years ago. First Nations peoples were the first to tell of how the land came to be, how the Rocky Mountains formed, and how the animals that live here first appeared. Scientists started telling their version of the story in the 1850s, when the British North America Exploring Expedition, under the command of Captain John Palliser, began charting the region's natural resources and its suitability for settlement.

Generations of geologists and palaeontologists followed, collecting information and marking their maps. Slowly, a new biography for Alberta emerged.

With every passing day, the story becomes more detailed. Scientists such as those who work at the Royal Tyrrell Museum of Palaeontology, in Drumheller, build upon and rework the theories, discoveries, and interpretations of previous years to answer today's questions. Researchers collect clues left by time in the rocky pages of the province's history. Impressions of 50-million-year-old leaves and flowers, burrows of 505-million-year-old soft-bodied invertebrates, three-toed footprints that measure a half-metre across, sandstone, mudstone, slate, and shale—these are some of the geological words and phrases that palaeontologists and geologists seek to decipher as they translate the past.

The history of the planet is divided into units of time. The two largest units are the Precambrian (Before Cambrian) and the Phanerozoic (All Life). Both are subdivided into smaller units: the Archean (Ancient) Eon and the Proterozoic (Former Life) Eon within the Precambrian, and the Palaeozoic (Early Life), Mesozoic (Middle Life), and Cenozoic (Recent Life) eras within the Phanerozoic. Eras are divided into periods, and periods are divided into smaller intervals of time.

Scientific research parallels the transformation of landscapes and human history through time: each is the sum of the past. Today's theories are the result of decades of scientific thought, accumulating evidence, and continual questioning—just as Albertans are shaped by events that took place yesterday, last year, and centuries ago, and just as present landscapes are built from the reworked remains of ancient environments.

A visit to Drumheller requires a journey deep into the past. As you descend the highway that cuts into the valley, you quickly pass the thin layer of topsoil that nourishes farm crops. The yellowish silts and gravels beneath the soils are debris carried to the area by glaciers 10,000 years ago. Immediately below is the Horseshoe Canyon Formation, a series of rocks that formed 72 million to 68 million years ago. These layers of sandstone, shale, and coal—remnants of ancient river deltas, backwater swamps, and floodplains—make up the valley's basin.

Outside the Royal Tyrrell Museum of Palaeontology, weather and water unmake the landscape. Seventy-million-year-old rocks dissolve, millimetre by millimetre. The sand, dust, and gravel wash into streams,

Dennis Braman, Royal Tyrrell Museum

Daishowa-Marubeni International Ltd.

Tim Schowalter

Alberta's landscape is a product of what has gone before. Long-ago events, plants, and animals formed the raw materials and modern environments for industry in the province today; petroleum (top), forestry (middle), and farming (bottom).

Royal Tyrrell Museum

Royal Tyrrell Museum

Royal Tyrrell Museum

ponds, and the river to collect on older sediment layers where the river flows into Hudson Bay, far to the northeast. Today, as happened millions of years ago, new rock forms from ruins of older rock. The progression of rock layers up the walls of the Red Deer River valley recounts the long story of the life, death, and rebirth of the local landscape through time.

Geological history is easy to follow in the badlands of the Red Deer River valley, but all of modern Alberta is similarly built upon reworkings of the past. The building blocks that make today possible were laid down millions of years ago. If plankton and stromatoporoids had not colonized Alberta's Devonian seas, the province would have no petroleum. Coal seams and salt beds that formed long ago are mainstays of many modern mining communities. Much of Alberta's agriculture industry relies on soils that collected as thousands of years passed. The grass that grows from these soils evolved about 55 million years ago from flowering plants, which appeared 100 million years earlier. Mixed woods of both flowering trees and conifers are the focus of provincial forestry. Visitors spend time and money in Alberta because of features such as the Rocky Mountains, which formed 35 million years ago from rocks that are as old as one billion years, and the remains of ancient animals preserved in the province's rocks.

All fossils in the province, including oysters (top), amber (middle), and dinosaur trackways (bottom) are protected by law. Excavation, trade, sale, or export of Alberta fossils and Alberta fossil products is illegal without prior permission from the Ministry of Community Development, which oversees the province's palaeontological resources.

chapter two

BEGINNINGS: THE PRECAMBRIAN ERA

This early Earth was nothing like the planet we know. It was hot. Air and water were acid. Land and oceans were constantly reshaping themselves. It was so hellish that scientists have named the geological period that spans Earth's birth and infancy the Hadean period for its resemblance to the steaming, fiery underworld of the ancient Greeks. Yet, despite its inhospitable nature, Earth in those early times was constrained by the same basic geological processes that occur within the Earth we know. Together, the planet's shifting, building land, oceans, and atmosphere set the stage for the next 4.6 billion years.

DIFFERENCES

The river's roar drowns out the cries of pelicans soaring and falling over rocky islands amidst the rapids.

Here at Mountain Rapids, the Slave River drops down granite staircases in one of four sets of rapids that lower the river more than 33 metres in the 27 kilometres between Fitzgerald, Alberta, and Fort Smith, Northwest Territories. The region's boreal forest and muskeg swamps are home to caribou, moose, deer, black bear, cougar, and wolverine. In forest clearings, cranberry, alder, and balsam poplar shelter songbirds that navigate the sky currents from Central America and South America every spring. White pelicans dive into the river, fishing for northern pike, whitefish, and goldeye.

Beneath this blanket of life lie the ancient, worn foundations of Alberta. These rocks are the roots of a mountain range that thrust towards the sky when primordial continents collided. Wind, water, and ice wore away the peaks, leaving only granite sutures where ancient landmasses were bound together. The scarred, rounded ledges are remnants of rock that formed almost three billion years ago—long before Alberta existed. At that time, Planet Earth looked nothing like it does today: land was barren and air poisonous. Life existed only in the form of single-celled, ocean-dwelling microbes.

Home to countless species of wildlife, including rare white pelicans (top), northeastern Alberta (middle) is one of the few areas in the province where Precambrian-aged rocks can be seen. At Mountain Rapids (bottom), the Slave River rushes over outcrops of these three-billion-year-old rocks.

OUT OF THE DUST: HADEAN EARTH

Long, long ago, in an arm of the spiral galaxy we call the Milky Way, a cloud of gas spun through space. Formed from remains of exploded stars, this nebula contained the seeds of our solar system.

About five billion years ago, the slowly spinning nebula spasmed. In the hot, dense centre, molecules were ripped apart and resorted by gravity. These nuclear reactions sparked our Sun. Swirling around the infant star, remaining gas flattened into a disk and condensed into tiny particles.

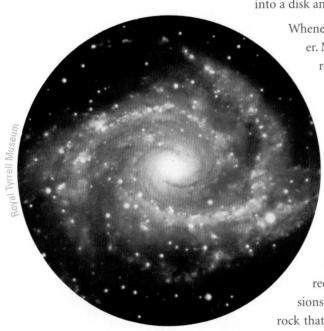

Royal Tyrrell Museum

Whenever they touched, the mineral grains stuck together. More and more particles bumped and bonded, and rocky bodies grew. As they attained mass, they exerted gravity of their own, attracting other objects to them. The number, frequency, and violence of impacts increased.

By 4.6 billion years ago, our planet Earth was born.

Shaped by violence and destruction, infant Earth was a dangerous place. Poisonous gases choked the atmosphere. Asteroids as big as mountains slammed into the planet, increasing its size and adding more minerals and gases. Like red-hot boulders dropped into cold water, the collisions evaporated oceans. They also melted much of the rock that formed Earth. Impacts kept the planet in flux as the growing amount of material that made up the planet was repeatedly melted and remixed. Scientists estimate that it took Earth about half a billion years to sweep its orbit of most dangerous asteroids and comets, and to leave its brutal beginnings behind.

Earth is located about two-thirds of the way down one of the Milky Way galaxy's spiral arms.

A thin crust congealed on the planet's surface. Today, after 4.5 billion years in which the planet has settled down, the crust varies in thickness from less than 10 kilometres below the deepest oceans to 90 kilometres beneath the tallest mountains. If the Earth were an apple, its crust would be thinner than the apple's peel.

Beneath the crust, giant currents began stirring within the thick, viscous layer called the mantle, tearing the fragile crust into plates, and pushing them against and apart from one another. At weak spots along the plates' edges, molten minerals poured out from the planet's belly to harden in layers. These were the first volcanoes. Their rocky piles collected along the edges of the plates as if they were flotsam over eddies in a river. Denser crust beneath oceans recycled back into the mantle under more buoyant continental plates. Water vapour released into the atmosphere by volcanoes and evaporating oceans condensed into torrential rains that filled ocean basins.

Even today, the atmosphere is the thinnest layer of the planet. Made of water vapour and gases that have escaped from the planet's interior by volcanoes, or were delivered to the planet by meteorites and comets, it is so sparse as to be barely breathable at the tops of the highest mountains. However, during Earth's earliest years, it was much thinner, and for most of the Precambrian period, humans would have found it unbreathable everywhere, as it contained almost no oxygen. The levels of oxygen we are familiar with today are the product of life forms that appeared only 2.5 billion years ago.

This early Earth was nothing like the planet we know. It was hot. Air and water were acid. Land and oceans were constantly reshaping themselves. It was so hellish that scientists have named the geological period that spans Earth's birth and infancy the Hadean period for its resemblance to the steaming, fiery underworld of the ancient Greeks. Yet, despite its inhospitable nature, Earth in those early times was constrained by the same basic geological processes that occur within the Earth we know. Together, the planet's shifting, building land, oceans, and atmosphere set the stage for the next 4.6 billion years.

PLUMBING EARTH'S INNERMOST SECRETS

Although humans can send spacecraft to explore space at the edges of our solar system, we have barely managed to scratch into the surface of the planet we call home. The deepest drill cores from mining and petroleum exploration wells come from only about 10 kilometres into the crust. Cores are samples of rocks that lie below the surface. They allow scientists to map the history and composition of the Earth's crust at that spot.

NASA/JPL/Caltech

We rely on indirect means to explore Earth's deeper interior. Seismic waves show the planet's internal structure and composition. By measuring how quickly the waves travel, scientists determine the density of what exists within Earth's crust, mantle, and even within the dense core at the planet's centre.

The slow, circling currents within the mantle also bring information to the surface. Magma and igneous rock tell scientists which minerals are common in the mantle, and what kinds of rocks line the volcanoes' routes through the crust.

Other sciences add to our knowledge of the Earth. Scientists have determined the age of the solar system by dating rocks gathered from the Moon's surface and meteorites that have fallen to the planet's surface. All formed approximately 4.6 billion years ago. By studying sister satellites within the solar system, we can take what we learn from looking at the sky and apply it to the planet that lies beneath our feet.

Because Earth's crust is constantly recycled, no rock survives from the planet's infancy. Scientists have determined the age of the solar system by dating rocks gathered from the Moon's surface and meteorites that have fallen to the planet's surface. All formed approximately 4.6 billion years ago.

Evolving Earth: The Precambrian

Earth's landmasses joined into supercontinents twice during the Precambrian Era: once about two billion years ago, and again about one billion years ago.

The rocks that make up Alberta come from many different places and have travelled great distances on the planet's continental rafts. Alberta's oldest pieces formed almost three billion years ago—long after life began.

During the hundreds of millions of years it took for the planet's crust to form and thicken, volcanoes spat out mantle material to add to Earth's first landmasses. Even though they are made of granites and basalts, continental rocks are more buoyant than the rock that lines ocean floors. Landmasses collected like floating debris on the planet's internal currents, growing as they collided and cemented together. Alberta's oldest rocks were scattered over three of these proto-continents. Parts of Alberta that now lie deep beneath Drumheller and Lake Athabasca formed tips of two continents. A segment of deep crust near Medicine Hat occupied a third landmass.

It took the Earth's inner currents to bring these pieces together. By about two billion years ago, all of Earth's proto-continents had shifted across the face of the planet to become the world's first supercontinent. Heat from the collisions welded some of the landmasses together. Limestone and shale that had formed on the underwater edges of the early continents were buried deep underground and pressure-cooked into other kinds of rocks or melted into magma.

But the surface of the Earth is never still. In a slow, perpetual dance, parts of Alberta have moved hundreds of kilometres over millions of years to join with other landmasses, to disengage, to add pieces to its collection, and to leave pieces behind. When the supercontinent that brought Alberta's first pieces together separated about 1.7 billion years ago, Alberta's rocky foundations remained behind. They formed the western section of the Canadian Shield, an ancient structure of igneous and metamorphic rock stretching across Canada and into the United States. Although exposed only in the northeastern corner of Alberta, the shield underlies the entire province.

Alberta lay next to an inland sea that gradually widened as landmasses drifted apart. The coast lay roughly along the line formed by the Rocky Mountains today. Nameless rivers meandered through granite hills, sandy plains, and coastal mudflats, collecting sediment and dumping it along Alberta's ancient coastline.

Life on land had not yet evolved, but it was abundant within the oceans. Cyanobacteria, a kind of microbe that produces its own food using sunlight energy, formed great reefs off the coast of Alberta. Levels of oxygen gradually increased in atmosphere and oceans.

During the Proterozoic Eon, Alberta's barren granite hills slowly eroded, their sediments washed by nameless rivers to the sea.

The continents joined a second time in the second part of the Precambrian, during the Proterozoic Eon, only to break up about 770 million years ago. Antarctica and Australia, two continents now in the Southern Hemisphere, floated towards the North Pole. An ancient ancestor to the Pacific Ocean flowed into the gap, to lap at Alberta's north-facing shores.

The series of glaciers that scientists think may have blanketed the planet during the late Precambrian added to the growing collection of sediments off the coast of Proterozoic Alberta. Ice sheets scoured the province, grinding down its granite hills, sweeping clean its plains, and carrying the debris—rocks ranging from grit to bus-sized boulders—on rivers of ice out over the frozen sea. When things heated up—again and again—the glaciers loosed their grips on the baggage and bombed the seafloor.

NATURE'S NURSERY: ARCHEAN EARTH

First Life

Hellish though early Earth was, it was a hothouse for life. The instability of the planet's surface, its flesh-charring temperatures, and caustic oceans and atmosphere would have killed many modern organisms. Billions of years ago, they nurtured and nourished life in its earliest stages.

Early Earth was so successful as a nursery that life was established within only a few hundred million years of the planet's formation.

Royal Tyrrell Museum

Uncertain evidence of Earth's earliest life is preserved within the Canadian Shield. Four-billion-year-old rock from the Northwest Territories is among the planet's most ancient rocks. It contains traces of carbon with what some scientists believe is a biological signature—carbon that could only have originated within a living organism. However, additional evidence is needed to prove first life.

The planet's earliest confirmable life is preserved in rocks that formed 3.5 billion years ago—one billion years after the birth of Earth. The fossils are cells of ancient cyanobacteria, single-celled organisms that use sunlight to create

The Royal Tyrrell Museum houses one of the Earth's most ancient rocks. Collected in the Northwest Territories, the sample of gneiss, a kind of metamorphic rock, is more than four-billion years old. Although other specimens of Canadian Shield rock contain traces of ancient organic carbon, this specimen does not.

food. Found living in many environments around the world today, cyanobacteria are hardy and resilient. Throughout their long history, they have adapted to many conditions. They thrive in acid hot springs, in alkaline lakes, and even in deserts that have seen no rain for more than a century. They live in intense sunlight and in near darkness. They weather temperatures that would fry or freeze most organisms. They endure extreme oxygen levels, ultraviolet radiation, and x-rays. They even survive nuclear-bomb explosions. They are superheroes of the microbial world.

They would have easily survived the conditions of early Earth.

But cyanobacteria are not Earth's earliest life forms. Bacteria are simple, single-celled organisms, lacking even cell nuclei, but even they are too complex to have emerged ready-made from Earth's early oceans. Simpler organisms must have existed, mutated, reproduced, and become extinct during countless ages before cyanobacteria appeared.

To determine what those organisms were and how they came to be, scientists are looking at what thrives in places similar to Earth's early, extreme environments. Many of the most primitive life forms known today prosper in places that most organisms shun. In acid solutions and searing temperatures of hot springs and deep-sea vents, hardy microbes harvest mineral compounds such as ammonia, sulphur, iron, or methane from the boiling chemical soups released through the Earth's crust, and transform them into food. The chemosynthetic organisms feed complex local food chains of tube worms, crabs, molluscs, and other animals.

Discovery of these communities has led some scientists to believe that the ancestor of every living thing found on Earth today came into existence in deep-sea hot springs. The planet's early oceans were broths of organic compounds such as methane, ammonia, cyanide, and carbon dioxide. Protected within tiny pits in rocks and in gooey muds near mineral vents, and subjected to acids, high temperatures, and intense pressures, the organic compounds somehow organized themselves into strands of self-replicating

Wil Andruschak

V. Tunnicliffe, University of Victoria

The most primitive life-forms known today require extreme heat to survive. This leads scientists to suspect that high-temperature environments were the source of Earth's first life forms. Hot springs, such as those found in Banff National Park (top), and mineral vents on ocean floors (bottom) replicate those early environments.

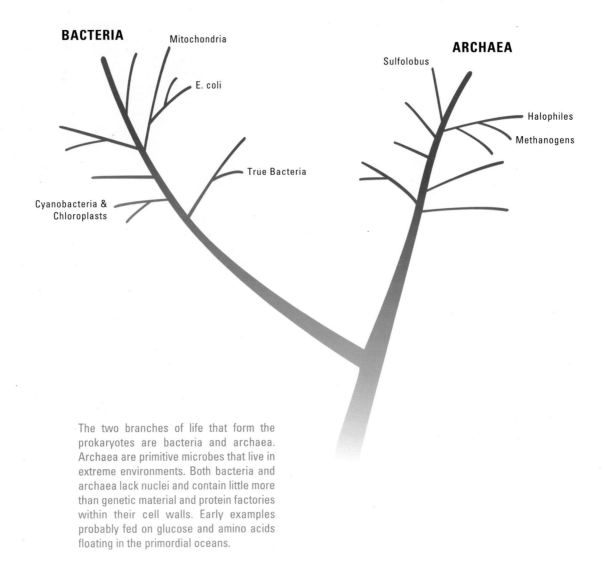

BACTERIA

Mitochondria

E. coli

True Bacteria

Cyanobacteria & Chloroplasts

ARCHAEA

Sulfolobus

Halophiles

Methanogens

The two branches of life that form the prokaryotes are bacteria and archaea. Archaea are primitive microbes that live in extreme environments. Both bacteria and archaea lack nuclei and contain little more than genetic material and protein factories within their cell walls. Early examples probably fed on glucose and amino acids floating in the primordial oceans.

proteins. The rocks and clays found near volcanoes and hot springs could have sorted organic molecules into the long carbon-based chains needed for life, encouraged formation of new, complex organic molecules, and even provided required minerals.

From self-replicating proteins, ribonucleic acid (RNA) evolved. As messengers that ferry or contain genetic information within cells, RNA is essential to all life on Earth.

Over ages, RNA lived, cloned, and mutated in the extreme nurseries where they formed. Without cell walls, these naked genes would have been vulnerable to the slightest changes in their environments. When some found protection within cell walls about 3.9 billion years ago, the prokaryotes, one of two major groups of life on Earth, were born.

WHERE IN ALBERTA

Alberta's rocky foundations are exposed in northeastern Alberta. The rounded, red granites found at Pelican Rapids, Mountain Rapids, and other places in the Slave River region formed as part of the Canadian Shield almost three billion years ago. As the planet's landmasses gathered into a single supercontinent, sedimentary shale and metamorphic rocks called gneiss were caught in the collision. Forced deep below the surface as the proto-continents ground into each other, the rocks melted and became magma.

When temperatures cooled, the magma crystallized into granite. Some slabs of original gneiss and sandstone survived, embedded within the outcrop.

Archie Landals

Alberta's only exposures of Precambrian aged Canadian Shield occur in the northeast corner of the province.

THE CHANGING PROTEROZOIC

Life in Abundance

Cyanobacteria were so adept at adapting to new environments that they quickly overran the planet.

In doing so, they altered the environment: they made it more favourable to themselves—and deadly to competitors. Cyanobacteria may have practiced an early, unintentional form of mass murder that cleared entire ecosystems of other living things, making room for future generations of the cyanobacterial kind.

The weapon of destruction was a byproduct of photosynthesis. In using sunlight to create food for themselves—an activity necessary to their survival—cyanobacteria filled the atmosphere and oceans with a chemical that, even today, is so corrosive that in high concentrations it dissolves flesh. It eats through cell walls and melts the delicate, life-giving structures inside.

Cyanobacteria were immune. They had evolved ways to repair cell damage caused not only by radiation and temperature but also by chemicals such as this. Many other early prokaryotes, however, were probably defenceless. When exposed to even minute quantities of the chemical, they would have leaked to death as if they were pierced balloons.

The agent of this chemical warfare was oxygen.

Early Earth was different from the Earth we know. Oxygen is essential for most modern life forms, but during the planet's first two billion years, there was little free oxygen in the atmosphere or oceans, and few life forms needed it. Concentrations of the gas likely never exceeded 2 percent, as oxygen atoms were easily

Royal Tyrrell Museum

Stromatolites can still be found in a few locations today. In places on the coast of western Australia and in the Persian Gulf, the living fossils are protected by high salt concentrations and lack of predators.

captured by other atoms such as carbon, sulphur, or iron to form other, more stable minerals.

The sunlit waters off the coast of Early Proterozoic Alberta and other landmasses were ideal habitat for photosynthesizing organisms. Cyanobacteria colonies sprouted up and formed reefs that ringed the continents.

These colonies, called stromatolites, were Proterozoic highrise apartment buildings. In the penthouses, exposed to sun and often to air, lived mats of photosynthesizing cyanobacteria. On floors below, photosynthesizing bacteria that could not withstand oxygen took refuge in the dim, low-oxygen shelter created by the high-living cyanobacteria. In the lightless basements, protected from sunlight and oxygen, light- and oxygen-evading bacteria fed on wastes produced by upstairs neighbours. Anytime a new floor was added, in the form of sand, silt, or mud washing over the topmost bacterial mat, the residents moved up a flight. As cyanobacteria raced upwards to claim the highest, sunniest spots on a mound, they cemented the sediment down with slime, creating lumpy, bulging, cabbagelike ramparts up to 10 metres tall.

At first, oxygen made by cyanobacteria had little effect on the planet's atmosphere. The gas, produced in the sunlit waters off the coasts of continents, bonded with iron molecules in the oceans. For almost two billion years, rust rained down from surface waters to the ocean floor.

Eventually, the Earth ran out of iron to absorb the oxygen. During the last two billion years, oxygen levels have steadily increased from 2 percent of the atmosphere to today's 21 percent.

Growing levels of oxygen forced vulnerable organisms to retreat to low-oxygen environments, where today many continue to thrive. Some organisms developed protective membranes. Others, including cyanobacteria, adapted to not only withstand the newly abundant chemical but also to take advantage of it: they used oxygen to turn food into energy.

With the increase in oxygen came the creation of the layer of the atmosphere that absorbs much of the Sun's deadly ultraviolet radiation—the ozone layer. This atmospheric sunscreen enabled life to colonize entirely new environments on the planet, such as shallow water and land.

Royal Tyrrell Museum

On the underwater margins of Archaean Earth's proto-continents, oxidized iron accumulated on the seafloor. Kilometres-thick banded-iron deposits, such as those found near Lake Superior, are the basis of the today's steel production industry. We owe our modern way of life to ancient cyanobacteria and the toxic byproduct of photosynthesis.

A New Kind of Life

Environments created by rising oxygen levels opened doors to new life forms. By 1.5 million years ago, a new kind of single-celled organism had emerged. Complete with a membrane-bound nucleus that protected its reproductive recipe, eukaryotes were larger, more complex, and better able to adapt to the newly oxygenated world than their more primitive prokaryote relatives were.

EUKARYOTA

Animals (including humans)

Plants

Fungi

Red Algae

Slime Molds

BACTERIA

Mitochondria

E. coli

ARCHAEA

Sulfolobus

True Bacteria

Halophiles

Methanogens

Cyanobacteria & Chloroplasts

Eukaryotes differed from their more primitive prokaryotic relatives by encasing their genetic information within protective cell nuclei. These first eukaryotes are ancestors to all modern multicellular organisms, including plants, fungi, and animals.

Some eukaryotes turned themselves into cooperative communes by hosting fellow organisms. In exchange for providing a protected environment to vulnerable guests, a host cell benefited from their labours. Photosynthesizing prokaryotes called chloroplasts produced food for the host; oxygen-breathing prokaryotes became mitochondria, the cellular power plants that turn food into energy; prokaryotes with tails that helped propel them through water loaned their use to eukaryote hosts.

Over time, the guest organisms became permanent residents—never checking out of the eukaryote hotel. The relationship was perpetuated, as most residents transferred their genetic information to the host nucleus. When the host cell replicated, the residents did as well—together, as a unit. Even mitochondria, which retain their own genetic coding, adapted the timing of their reproduction to that of host cells.

The evolutionary success of eukaryotes is due to the method they evolved to speed up genetic adaptation. Eukaryotes create genetic prototypes with every offspring. They do this by unzipping the genes of one parent and recombining them with those of a second parent. The results are similar but unique combinations of genes impossible to achieve otherwise. Some gene combinations better suit the environments the new individuals live in than others. The individuals with favourable combinations are more likely to survive and reproduce within that environment.

This method of gene recombination is called sexual reproduction. Its genetic mixing allows both rapid change within species and accelerated evolution of new species—an ideal strategy for dealing with uncertain, changeable environments.

There is little fossil evidence for much evolutionary experimentation during the first billion years of eukaryote existence. Most eukaryotes during this period were single-celled organisms such as algae, paramecia, and amoebae. However, some scientists suspect that during this time the groundwork for all body plans that exist today was being laid. These early innovations permitted later development of muscle, shell, and skeletal structures for feeding and moving, and allowed animals to diversify and spread to new ecological niches.

Among the new creatures were seaweeds and wormlike animals. Strange organisms that resembled air mattresses and sea-jellies also existed, living, as most Proterozoic organisms did, by absorbing food and gases from seawater. Although some larger, multicellular organisms did evolve, environmental stress was required to jumpstart diversification of life on Earth.

Swiftly Changing Planet

The second Proterozoic supercontinent split apart about 770 million years ago, and shallow, sunlit seas filling the continental rifts became perfect habitat for stromatolite-building cyanobacteria. Their reefs ringed North America and provided shelter to some of Alberta's earliest multicellular life.

The tropical idyll was short. Throughout the Rocky Mountains and on almost every continent, unusual juxtapositions of different rocks record what may be the most dramatic series of climate shifts Earth has experienced. Given that the continents rested near the equator at the time when the rocks formed, scientists have long been mystified about how layers of ancient glacial debris came to alternate with layers of carbonate rock, which forms only in warm water.

One theory suggests that Earth's entire surface repeatedly froze and thawed, beginning about 750 million years ago.

Carbon dioxide is a greenhouse gas; it traps heat within the atmosphere. According to the Snowball Earth theory, levels of this gas in the atmosphere dropped towards the end of the Proterozoic, causing temperatures to fall. Ice packs formed over the planet's poles. White ice reflects more solar energy than dark, open water: temperatures plummeted further, spurring a runaway climate-feedback loop.

Within 1,000 years, equatorial oceans were buried under ice up to one kilometre thick. Temperatures settled around –50° Celsius.

From space, Earth would have looked like a giant snowball: no blue water, no drifting clouds, no green and brown continents . . . just white ice and an occasional volcano thrusting through to belch its brew of gas and magma from the planet's warm, restless belly.

As ice cut off sunlight and oxygen to the oceans, life would have suffered. Organisms requiring sunlight to survive were restricted to above-ice volcanoes or were cast adrift on snow and dust embedded in ice. Seafloor hot springs were another haven; microbes feeding on mineral broths released from deep within the Earth's crust were the base of complex food chains. Cold-loving organisms, such as those that thrive in today's Antarctic, also would have survived.

Royal Tyrrell Museum

Alberta's oldest fossils are stromatolites found in the Waterton Lakes National Park region. These specimens lived during the Late Proterozoic.

In these isolated environments, stressed by the sudden shift in climate, new life forms would have quickly evolved to better withstand the changing environment.

According to the Snowball Earth theory, this first global deep freeze lasted 10 million years.

Volcanoes freed the planet from frostbite. Carbon dioxide released by volcanoes accumulated in the atmosphere, gradually climbing to levels 1,000 times greater than today. The atmosphere became the greenhouse that thawed the planet. When ice along Alberta's ancient tropical coastline melted, it loosed its payload of pebbles, rocks, and boulders carried from the province's interior. These bombarded the seafloor, where they formed distinctive layers of glacier-fed rock.

Within centuries, Earth's climate swung from being an icehouse to a sauna. Temperatures soared to 50° Celsius. Thick blankets of calcium-carbonate muds piled up on ocean floors— lasting signposts of high temperatures and extreme carbon dioxide levels.

As conditions for life improved, life forms that had evolved in protected hideaways emerged to repopulate the globe.

But the challenges were not over. Alternating glacial–tropical rock layers found on every continent today indicate Earth experienced as many as four icehouse–sauna episodes between 750 million and 580 million years ago.

If Earth did alternate between global glacier and worldwide tropics during the late Proterozoic, as the Snowball Earth theory proposes, these rigorous hot-and-cold treatments may have been the evolutionary cleansers and fortifiers that stimulated Precambrian life, clearing the way for complex animal life that was to follow.

Continental Shuffle

Over the course of Earth's history, continents have merged and separated repeatedly. Floating on top of the slowly churning mantle, the shuffling of the 15 or so crustal plates across the planet's surface is called plate tectonics.

Continental crust is rarely recycled into the mantle; it is too buoyant to be sucked down where crustal plates collide. By contrast, the heavy crust that floors the oceans is recycled constantly, with only occasional pieces forced aboard floating continental barges. The oldest oceanic crust is only two billion years old.

New ocean crust is built along midocean ridges, where underwater volcanoes disgorge mantle material onto the seafloor, forcing older, hardened crust material to spread away from the ridges. The edges of the ocean plates nudge adjoining plates, gradually rearranging the surface of the planet. The Atlantic Ocean is currently growing, pushing the North American and European plates apart. Our continent travels a few centimetres every year—about the length a fingernail grows in one year.

As North America is slowly shoved westwards, it encounters the plates of the Pacific Ocean. With no place to go but down, the ocean plates are forced under North America—again, at a speed of centimetres each year. At first, the impact crumpled the prow of the North American plate—a head-on collision in very, very slow motion—forming long, parallel mountain ranges from today's coast to the Rocky Mountains. Friction caused by one plate grinding past the other is the source of western North America's earthquakes. Its volcanoes are the result of molten material being squeezed upwards, like toothpaste,

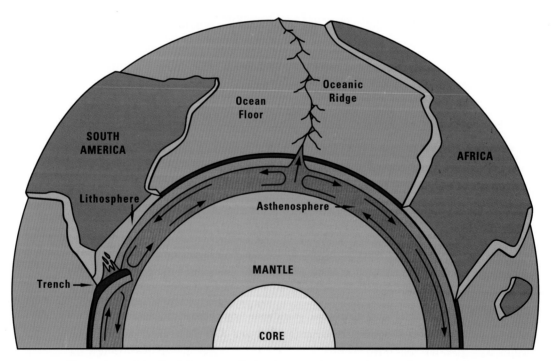

The surface of the Earth is never still. Continents and ocean plates nudge each other, changing the planet's face over millions of years and causing volcanoes and earthquakes.

WHERE IN ALBERTA

The Purcell Supergroup, in Waterton Lakes National Park, documents the coastal environment of the province's earliest days. The 11-kilometre-thick collection of rocks contains layers of sandstone, slate, shale, and limestone—all sedimentary rocks that formed in water. This is one of few places where sedimentary rock of this age remains unaltered by metamorphism. The striking red colour of many of the area's mountains is a result of oxidation of iron contained in the ancient mudflats. Green-coloured rock indicates where the mudflats were not exposed to air, possibly because the mud collected in deep water.

Cliffs lining many of the region's creeks tell more of the story. The rocks along Cameron Falls formed as part of the shallow-water mudflats that lay off the coast of the early North American continent 1.5 billion years ago. They contain fossilized stromatolites, the remains of cyanobacteria colonies. Along Drywood Creek, north of the park, and in Red Rock Canyon, changes in ancient water levels are recorded as ripple marks, mud cracks, and moulds of salt crystals in the rock.

The vivid red of ridges and peaks throughout the Waterton park region (top) results from the oxidation of iron contained in ancient mudflats. Rocks at Red Rock Canyon, Drywood Creek, and other places throughout the region preserve fossil stromatolites at Cameron Falls (middle) and ancient ripple marks (bottom).

Dennis Braman, Royal Tyrrell Museum

©Parks Canada/Simon Lunn

Archie Landals

WHERE IN ALBERTA

The volcanic rocks that pierce the Purcell Supergroup formed about 1.1 billion years ago. Pillow lavas on Table Mountain show where lava poured out of the ground onto the ancient seafloor.

A collection of rock formations found along the length of the Rocky Mountains recalls the time when ancient Alberta froze and thawed with the rest of the planet. The Windermere Supergroup contains layers of coarse sandstone, pebble conglomerate, crumbly slate, and carbonate limestone—testament to the alternating presence of ice and tropics. Easily eroded compared to later rocks, these often form the lower, more rounded slopes of the Rocky Mountain Main Ranges.

In places, layers of rock representing the Late Proterozoic are nine kilometres thick.

Tim Schowalter

Mike Todor

Wil Andruschak

Pillow lavas (top) pierced the Purcell Supergroup about 1.1 billion years ago.

Evidence for the Snowball Earth theory can be found in exposures of the Windermere Supergroup (middle) along the length of Alberta's southern Rocky Mountains, including the lower slopes of Storm Mountain (bottom).

chapter three

THE LIVING SEA: THE EARLY PALAEOZOIC ERA

According to the fossil record, fully developed multicellular animals appeared suddenly and in great numbers 545 million years ago, marking the beginning of the Cambrian Period. Most of the creatures we know from that time had hard, mineral shells, spines or sheathes, which resist decay. Scientists coined the term "Cambrian Explosion" more than a century ago to describe this apparent abrupt proliferation of animals.

CAMBRIAN EXPLOSION

July 1999

C.J. Collom

Castle Mountain, one of Banff National Park's most photographed peaks, is one of the few places in Alberta where half-billion-year-old rock can be seen. On the slopes of Mounts Field and Stephen, in nearby Yoho National Park, similar rocks contain some of Earth's oldest and most famous animal fossils: these are the Burgess Shale and Mount Stephen trilobite beds.

Palaeontologist Christopher Collom and colleagues climbed Castle Mountain's cliffs in 1999. Their destination: the Stephen Formation, a crumbly, 40-degree shale slope sandwiched between layer-caked limestone platforms known as the Cathedral and Eldon formations.

Within minutes of hauling himself up to the shale that day, Collom recognized the importance of the site. Preserved in the grey rock are tracks, trails, and burrows of ancient animals—505-million-year-old trilobites and wormlike creatures. They are similar to the animals preserved in the Burgess Shale and Mount Stephen trilobite beds.

"We were probably the first geologists to set foot there since the 1940s," recalls Collom. "The amount of impressive geological and palaeontological material lying about was mind-boggling!"

The Stephen Formation is a 40-degree shale slope, sandwiched between limestone cliffs on Banff National Park's Castle Mountain (top). In 1999, geologist Christopher Collom and colleague Arden Bashforth (shown) scaled Castle Mountain to examine the exposure. They discovered half-billion-year-old trace fossils, including trilobite trackways and worm trails (opposite page, top).

Trace fossils such as those found on the narrow slopes of Castle Mountain are preserved behaviours of ancient animals, shadows imprinted in stone long after the creatures that made them became extinct. Discoveries such as those made at Castle Mountain help Collom, who teaches at Calgary's Mount Royal College, and research partner Paul Johnston, Curator of Invertebrates at the Royal Tyrrell Museum of Palaeontology, to answer questions about how early animals evolved, interacted, and survived.

Palaeontologists have been asking those questions since the Burgess Shale was discovered a century ago.

C.J. Collom

C.J. Collom

A Cambrian Community Revealed

Courtesy of Smithsonian Institution Archives

Palaeontologist Charles D. Walcott discovered the Burgess Shale fossils at summer's end in 1909. Lured to the area by the famous Mount Stephen Trilobite Beds which rest on a mountainside to the south, he and his family had spent the season searching the lofty peaks and ridges of Canada's Yoho National Park for fossils of animals that lived during the Cambrian Period, 545 million to 490 million years ago.

On a trail across the southwest flank of Mount Field, high above Emerald Lake, Mrs. Walcott's horse stopped. A large rock had slid from higher up the slope to block the trail. Walcott dismounted from his own horse to move the slab. As he bent down to shove it off the trail, he noticed a peculiar, dark grey smudge along the rock's broken edge. It was a fossil of an animal he had never before seen. When Walcott split the slab open, more fossils were revealed.

During nine expeditions, Charles D. Walcott (above) collected more than 65,000 specimens representing 150 species from the Burgess Shale for the Smithsonian Institution in Washington.

Construction workers building the railroad through the Canadian Rockies discovered the Mount Stephen Trilobite Beds in the late 1800s, in what is now Yoho National Park. The site contains thousands of fossils of trilobite *Ogyopsis* (right).

Paul A. Johnston, Royal Tyrrell Museum

Royal Tyrrell Museum

Known from more than 15,000 specimens collected from Walcott's quarry, *Marrella* is the most abundant Burgess Shale animal. It is a distant relative of lobsters, spiders, and trilobites.

Royal Tyrrell Museum

High on Fossil Ridge beneath Mount Wapta, the field season is short. Researchers usually shovel snow from the quarries in mid-July and frequently face end-of-season blizzards in late August.

As director of the Smithsonian Institution in Washington, DC, Walcott was an accomplished palaeontologist. Even in 1907, when he first came to the Canadian Rockies to study the Mount Stephen Trilobite Beds, he was famous for his research into Cambrian-aged fossils. When he saw the ancient remains in the shale that August day, he immediately recognized the importance of the specimen and the slope where it was found.

The family set up camp. They collected more fossils and tried to pinpoint their source on the rocky slope stretching 200 metres above the trail. Within days, Walcott and his family had located the general area of the fossil beds and collected many slabs containing diverse and well-preserved specimens. What really excited Walcott, however, were the fossils of soft-bodied animals.

Walcott was aware that the fossil record favours animals with hard body parts over animals with only soft tissues. When an animal dies, its body decays. Muscles, nerves, skin, and other soft tissues rot quickly; hard body parts such as bone and shell take longer to disintegrate. This increases the chance they will remain intact long enough to be buried and preserved. More animals with hard body parts are fossilized than are those without. The bias in preservation affects our knowledge and understanding of early animal communities.

According to the fossil record, fully developed multicellular animals appeared suddenly and in great numbers 545 million years ago, marking the beginning of the Cambrian Period. Most of the creatures we know from that time had hard, mineral shells, spines or sheaths, which resist decay. Scientists coined the term "Cambrian Explosion" more than a century ago to describe this apparent abrupt proliferation of animals. Since then, however, discoveries of trace fossils show that complex animals had existed farther back in time, during the latter part of the Precambrian Era. Gene-sequencing research suggests their first appearance may have occurred as far back as one billion years.

The fossils that Walcott and his family discovered on the side of Mount Field that summer included a rare complete community of ancient animals. Not only were fossils of shelled creatures such as trilobites and brachiopods found intact, but also preserved were remains of worms, sponges, and other soft-bodied animals. In some specimens, the preservation is so fine that the animals' last meals are seen within fossilized guts.

The Walcott family delayed their return to the nearby town of Field for a week, but the increasing risk of being trapped in the high country by late-summer blizzards forced them off the mountain. They returned to Washington in early September. Walcott spent the winter attending to his Smithsonian duties—and wondering, hoping, and planning for the next summer. He had ben so close to pinpointing the place on the mountainside that was the source of the amazing fossils . . . and now he had to wait 10 months to finish the task.

In July, Walcott was back in Field, armed and prepared for a final assault on the fossil beds. However, in the high country of the Canadian Rockies, winter imprisons the land until halfway through summer: the fossils were buried under metres of snow. The family waited in town until the end of the month to get near the fossils. Once the snow cleared, they climbed over the mountain and set up camp. Within days, they had mapped and measured the exposed section of the Burgess Shale.

On August 2, Walcott located the main fossil-producing bed, now called the Walcott Quarry.

The Burgess Shale and its spectacular surroundings drew Walcott and his family back to the Rockies summer after summer. Through the years, they collected more than 65,000 specimens from the mountainside for the Smithsonian.

Many of the fossils were of creatures unlike anything Walcott and his colleagues had seen before. More than 150 different species are represented. Some belong to modern groups of related animals, but some represent previously unknown, now-extinct phyla. Better than any site known previously, the Burgess Shale documents the Cambrian Explosion, a time when early animal life was quickly diversifying.

Walcott classified Hallucigenia as a bristle worm; however, it is now considered to be a velvet worm.

Bristle worms live on modern ocean floors.

Carnivorous worms related to ancient Ottoia are rare in today's oceans.

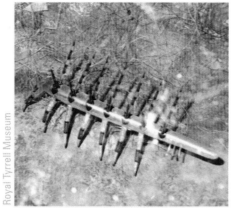

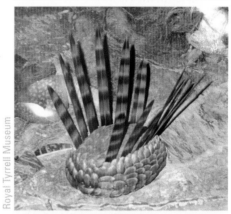

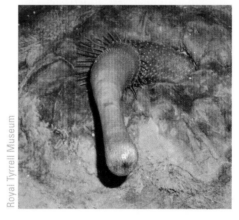

Royal Tyrrell Museum

Hallucigenia (top) was thought to have walked on two rows of spines, which line one side of the animal's body. Fossils from China and preparation of other specimens indicate the long-held view of *Hallucigenia* was upside down and backwards. Its spines protected the creature's back.

Sluglike *Wiwaxia* (middle) crept along the seafloor, its five-centimetre-long body armoured in thousands of spines and tiny scales. Walcott considered *Wiwaxia* a bristle worm, but subsequent research suggests the animal belongs in its own distinct phylum.

Ottoia (bottom) is the most abundant Burgess Shale worm. It grasped prey with spines on its extended proboscis and mouth, and pulled them into its trunk. Specimens with preserved guts show that it ate *Haplophrentis* and other *Ottoia*.

At the start of the Cambrian Period, 545 million years ago, what is now the spine of North America—the Rocky Mountains—was beachfront on the continent's northern coast. Alberta basked in sunshine near the equator, its shores lapped by a precursor to the Pacific Ocean.

For the first time in hundreds of millions of years, North America floated free from other continents. This allowed its edges to slowly sink. As Alberta eroded and flattened out, lifeless rivers carried its sediments to the seashore. The ocean slowly moved inland, marking the first of many repeated floods and retreats during the course of more than 180 million years. Soggy periods were relieved by dry spells, but each time the sea returned, it covered more of the continent.

By the end of the Cambrian, 490 million years ago, all of southern Alberta and parts of Saskatchewan and Manitoba were underwater. During the Ordovician Period, the north end of the continent began to grind into Greenland and Europe, lifting the land and gently tipping the seas out of western Canada. However, by the end of the Ordovician, the ocean once again drowned most of North America. Reef builders re-established their rocky homes on the underwater edges of the continents— laboratories for the evolution of animal communities.

A second collision between North America, Greenland, and Europe emptied the seas from Alberta at the end of the Silurian Period.

Evolving Earth: Early Palaeozoic

During the early Palaeozoic Era, Alberta faced north-ward near the equator, along the coast of what is now North America. Alberta, 545 million years ago.

Wide rivers crossed Alberta during the Cambrian Period, carrying sediment to the sea. Underwater reefs on the continent's supported evolving communities of complex animals.

Alberta's Cambrian seas were warm and shallow—only a few hundred metres deep. Fine, white, limy muds, like those that tint the waters of the Caribbean today, covered the seafloor, drifting into mounds and shoals. On the edges of the flooded continent, algae and invertebrates built massive limestone ramparts. Some of these reefs extended more than 20 kilometres along the Alberta's edge—tiny sections of a 2,000-kilometre reef structure that ringed the continent. During drops in sea level, silts and clays washed from newly exposed land buried reefs and shoals.

In this world of sun-dappled waves, shifting mud, and rocky reefs, animals evolved from minuscule mud dwellers into complex creatures. Oxygen concentrations in the atmosphere and ocean reached a then-whopping 10 percent, fuelling increased activity among oxygen breathers. Animals developed shells, spines, and sheaths—and new ways of living. No longer were animals restricted to the role of seafloor garbage collectors in their search for food— many now turned their hungry gazes upon each other.

Predators were on the hunt for fresh invertebrate flesh.

This new lifestyle unleashed an ever-escalating evolutionary arms race. To avoid being eaten, animals developed personal fortresses of shells, spines, exoskeletons, and tubes to protect their vulnerable bodies. In response, predators evolved claws and other hard grasping appendages to overcome armoured victims.

The innovations served as more than weapons and aids for self-defence. Hard body parts help animals move and feed by providing rigid supports and braces for muscles and other soft tissues to pull against. The new body designs made possible the evolution of many later groups of animals, including vertebrates with their backbones and bony internal skeletons.

Royal Tyrrell Museum

Royal Tyrrell Museum

Royal Tyrrell Museum

Anomalocaris was a formidable predator of the Burgess Shale. Twenty-centimetre-long limbs, attached to the front of its head, helped this sea monster capture *Canadapsis* and other prey. *Anomalocaris* fossils have been found in North America, China, Australia, and Greenland.

Opabinia had five eyes at the front of its head and a vacuum-hose nose that ended in grasping spines. This unique feature may have been used to grab prey, such as *Yohoia*, or to extract worms from burrows.

Tiny *Pikaia's* body is supported by a rod to which muscles attached, making it one of the earliest-known chordates—among our first-known ancestors. The tail flares into a fin, which would have helped the animal propel itself through the water just above the seafloor.

On the Edge

Paul A. Johnston, Royal Tyrrell Museum

The Cathedral Escarpment stretches through more than 100 kilometres of Canada's Rocky Mountains, forming cliffs on a few peaks, then reappearing several mountains farther south. These outcrops are relics of the underwater edge of Cambrian North America. Scientists who study the early evolution of animals search the shale that lies next to the limestone cliff for remains of creatures that lived during the Cambrian Period—creatures such as the animals of the Burgess Shale.

About one-half-billion years ago, limestone reefs bounded a warm, shallow-water shelf that stretched off the coast of Cambrian North America. On the shallow, sunlit plain between reef rim and shoreline, limy muds piled up, moved about by tides and currents, and battered by waves and erosion as water level rose and sank. In the other direction, dark shales sloped hundreds of metres from reeftop to ocean floor.

The edge of the limestone shelf was unstable. The rim frequently failed, tumbling reef and shale into deeper waters and leaving a sheer limestone cliff rising from the seafloor. This cliff is the Cathedral Escarpment of the Cambrian Period. As time passed, layers of mud collected at the cliff's foot.

The ancient animal communities that we now call the Burgess Shale lived in and above the mud, in the shadow of the Cathedral Escarpment. Animals such as cone-shaped Haplophrentis may have fed on the mud, eating it for the tiny food particles trapped between grains of sediment, or grazed on bacteria floating above the seafloor. Mud dwellers and bacteria grazers were food for other animals: predators such as Ottoia and Anomalocaris were well equipped to hunt and capture both armoured and soft-bodied prey.

Exposures of the Cathedral Escarpment can be found in sections along more than 100 kilometres of the Rocky Mountains. It is a relic of what was once the edge of North America.

Deep water protected the communities. Storms taking place in shallow water would scarcely have been felt in the community below. However, storm waves that crashed into the escarpment may have added to the cliff's instability. Life in the eat-or-be-eaten Burgess Shale was frequently interrupted. Boulders broke off the cliff and careened into the deep muds. Swirling, suffocating sediments swept over the cliff, burying the Burgess Shale animals. Black muds covered the animals; dark, oozing clays slowed decay of their bodies. As community after community was smothered, newer communities of Cambrian creatures colonized the muds. Slowly, over time, the muds and clays at the foot of the cliff piled higher, pressing on lower layers, squeezing water out, and compressing the clays and muds into rock

Royal Tyrrell Museum

The Royal Tyrrell Museum's Burgess Shale exhibit provides an enlarged glimpse of an early, underwater ecosystem as it may have existed 505 million years ago. Models in the exhibit are 12-times life size, and they represent 45 Burgess Shale species.

LAND AND SEA: THE ORDOVICIAN AND SILURIAN PERIODS

Terrestrial Pioneers

Scientists had long believed that plants colonized land before animals left the sea. However, a 2002 discovery in eastern Ontario suggests that animals took their first steps on land 40 million years earlier than had been thought. Sometime between 480 and 500 million years ago, during the first few million years of the Ordovician Period and possibly as far back as the Cambrian Period, invertebrate animals crawled from the sea and left tracks in sand dunes on a beach. More than 25 of the ancient trackways are preserved in an abandoned sandstone quarry north of Kingston. They were made by long-tailed, eight-legged arthropods which probably left the sea only for limited periods of time.

The migration of plants from scum-covered coastlines and ponds to rocky shores is believed to have begun about 470 million years ago. Although scientists do not know how the earliest land plants looked, they may have been a kind of alga or primitive moss. Early green pioneers maintained close ties to water: dehydration, gravity, and limited nutrient absorption made survival on dry land more difficult than living in water.

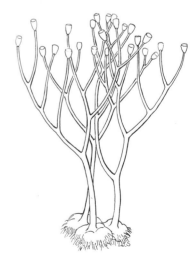

The oldest known vascular plant, *Cooksonia*, consisted of stalks topped by spore-filled sacks. It stood only a few centimetres tall and lacked leaves and roots.

TRILOBITE LURES

Royal Tyrrell Museum

Olenoides is one of the Burgess Shale's trilobites. Exceptional preservation at the site shows not only the three parts of the animal's exoskeleton, but limbs and antennae as well.

The siren song of scientific discovery has long lured palaeontologists to the Mount Stephen Trilobite Beds above Field, British Columbia. The tomb of strange, ancient creatures drew Charles D. Walcott to the Rockies in 1907 on a quest for fossils that lived during the Cambrian Period. He had spent his life studying trilobites and other early animals, and could not resist the lure of discovering other, similar caches of trilobite remains.

Trilobites first appeared about 525 million years ago, when the supercontinent that had formed during the Late Precambrian was drifting apart. The animals burrowed in seafloor muds and swam in shallow reef waters for more than 300 million years. The proximity of major landmasses to one another aided the spread and diversification of trilobites, as their fossils are found on almost every continent. They are as small as one centimetre in length, and as large as 75 centimetres, as documented by a 445-million-year-old specimen discovered near Churchill, Manitoba, in 1998.

Trilobite body design is distinctive. Three side-by-side lobes that form the body give the animals their name. They walked on jointed legs which bent in the same manner as the arms of folding desk lamps. As arthropods such as lobsters, crabs, insects, and spiders do, trilobites wore their skeletons on the outsides of their bodies—combination suits of armour for protection and braces for muscles. As they outgrew the skeletons, trilobites would discard the old, hard carapaces and grow larger skeletons—a process repeated many times throughout an animal's life.

TRILOBITE LURES

Most trilobite fossils are discarded exoskeletons. Palaeontologist Brian Chatterton, of the University of Alberta, describes a 430-million-year-old Royal Tyrrell Museum fossil from Quebec as an exoskeleton conga line. The specimen is a fossilized burrow that contains discarded skeletons of 27 individual trilobites, packed together, piled on top of one another, head to toe, back to back. Some are lined up as if they were dancing the conga.

"These animals were hiding together in a hole while they were going through their moult stage," says Chatterton. "They would have been very vulnerable throughout the time it took for their new carapaces to form. They would not have been able to move very fast, because many of the muscles they used to move around anchored to the exoskeleton. And on top of that, without their carapaces, they lacked the armour that normally protected them from all kinds of predators."

Nearly a century after Charles Doolittle Walcott explored the Mount Stephen Trilobite Beds, the fascination for trilobites continues: they are among the best-known and most-studied early animals with hard body parts. Chatterton, who has studied trilobites for more than 30 years, says their allure is in the way they capture the imagination. "Trilobites are fun animals: they moved around; they did things; they had all kinds of really cool life cycles that we can track. So many different kinds of trilobites evolved over such a long period of time, and each kind lived its life in it own distinct trilobite way."

Brain Chatterton

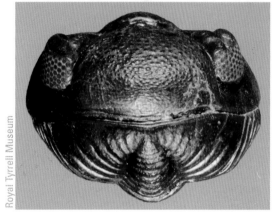

Royal Tyrrell Museum

The 27 trilobites (top) that left their outgrown exoskeletons in this 430-million-year-old burrow ranged in size from juvenile to adult, and may provide evidence that males and females of this species looked different from one another. The specimen also helps prove that some trilobites were opportunists. They took advantage of whatever hiding holes were available in an area to protect themselves during vulnerable moult stages.

Trilobites are the first animals known to have evolved complex eyes (bottom). The ability to spot prey on the seafloor or in the water column and to see predators approaching likely added to trilobites' success as a group.

SOLVING ANCIENT MYSTERIES

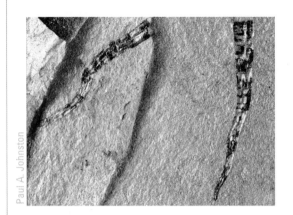

Paul A. Johnston

Worm tube *Byronia* is found in abundance in pockets along the Cathedral Escarpment. Its presence at Miller Pass led scientists to suspect ancient underwater hot springs and mineral seeps played a role in Burgess Shale–like sites. Specimens of *Ogyopsis* at the Mount Stephen Trilobite Beds occur with pyrite, one of the sulphide minerals found along the escarpment.

A need for fossils for an exhibit at the Royal Tyrrell Museum led to a new, ongoing research project for palaeontologists Paul Johnston and Christopher Collom. The Museum was developing its exhibit on the Burgess Shale in 1998 and needed to expand its collection of Cambrian-aged fossils. As it was not possible for the museum to collect fossils from the Burgess Shale itself, Museum palaeontologist Paul Johnston and Mount Royal College palaeontologist Christopher Collom searched out similar sites in the southern Rockies.

They found one during their first expedition. Preserved within rock high on the slopes above Miller Pass in southeastern British Columbia were the remains of trilobites *Naraoia* and *Hemirhodon*, worm tube *Byronia*, alga *Margaretia*, and ice-cream-cone-shaped *Haplophrentis*. The rock in which the fossils were found forms part of the Duchesnay Formation, which is about five million years younger than the Burgess Shale. As at the Burgess Shale, the small pockets of Miller Pass fossils are located next to the preserved remains of the Cathedral Escarpment.

The scientists had recently finished researching fossil evidence of ancient hydrothermal seeps and chemosynthetic communities in northeastern Alberta. Fossil distribution and abundant sulphide deposits at Miller Pass prompted Johnston and Collom to speculate that the area may have also once been the location of undersea hot springs, or hydrothermal seeps. By feeding colonies of sulphur-digesting bacteria that attract other animals to the buffet, hot springs or seeps in modern seas encourage localized concentrations of animals. The scientists were especially intrigued when Johnston recalled that abundant fossils of *Byronia* and *Ogyopsis* also occur in pockets at the Mount Stephen Trilobite Beds.

The scientists had to wait until the following summer to visit Mount Stephen to test their hypothesis. Once on the mountain, however, they immediately noticed the giant, pillow-shaped carbonate rocks that are embedded within the dark shale. These rocks had long been thought to be blocks of limestone that, 505 million years ago, had broken off the nearby Cathedral Escarpment and fallen into the deep water below the cliff, where they became lodged in the Burgess Shale muds. But on this expedition, Johnston noticed that the rocks are rounded, not angular as broken bits of cliff should be. He quickly realized that they were preserved mounds of ancient carbonate mud—mud that had formed where cracks rising from deep within the crust released mineral-rich water along the escarpment. Chemical reactions and bacteria around many hydrothermal seeps today cause calcium-carbonate mud to pile up in mounds

SOLVING ANCIENT MYSTERIES

As the field season continued and Johnston and Collom broadened their search for ancient hot spring evidence, they found other similar mounds at other sites in the Burgess Shale.

When the scientists returned to Mount Stephen in 2001, they discovered an exposure of the Cathedral Escarpment up the slope from the trilobite beds which shows evidence that the cliff had once been exposed to super-hot, super-salty water which had steamed the limestone into dolomite. Immediately next to the ancient cliff, sulphide minerals and trilobite fossils are abundant.

The scientists now suspect ancient, underwater mineral seeps may be vital but overlooked features at Burgess Shale–like sites.

Paul A. Johnston

Indicators of ancient hydrothermal activity, carbonate-mud mounds near the Mount Stephen Trilobite Beds are massive—some more than seven metres high. Scientists have found similar, smaller mounds elsewhere along the Cathedral Escarpment.

It took tens of millions of years for successful solutions to these problems to emerge. The oldest known vascular plant dates from Silurian-aged rocks in Europe, about 415 million years ago. *Cooksonia's* short stalks contained veins which allowed the plant to transport water and nutrients soaked up from the ground to its tissues. Its stalk reduced water loss to the atmosphere. A primitive, internal woody structure permitted *Cooksonia* to stand upright.

By the end of the Silurian Period, 415 million years ago, green pioneers only slightly more developed than *Cooksonia* covered much of Earth's land.

These first plant and animal explorers were quickly followed by full-time settlers. Scorpions and millipedes live in deserts and forests today, but they are thought to be among the most ancient land-dwelling animals. Millipede burrows are known from Ordovician-aged fossil soils in Pennsylvania. Fossils of both scorpions and millipedes have been found in Silurian rocks in Scotland.

Today, known species of land-dwelling invertebrates, which include insects, arthropods such as spiders and scorpions, and millipedes, outnumber species of all other kinds of terrestrial animals.

Backbones and Jaws

The vertebrate descendants of primitive chordate animals such as *Pikaia* emerged during the Ordovician Period, 490 to 445 million years ago. Fishes are thought to be the first true backboned animals. The earliest fishes possessed only tail fins and looked like small, flexible torpedoes with head shields and scales. Lacking jaws, they filtered food through their mouth and gill openings. They probably spent their lives on the seafloor, grazing on mud for food particles. Later examples evolved fins and powerful tails, and may have lived among seaweed forests or rocky outcrops above the seafloor, returning to the bottom to feed.

Mark V. H. Wilson

Agnathans were jawless fishes that filtered food through their mouths and gill openings. Some, such as *Dinaspidella elizabethae*, had armour to protect them from predators.

Jawless fishes prospered for tens of millions of years, thriving in warm, tropical waters.

About 410 million years ago, at about the time when a second collision between North America, Greenland, and Europe emptied the seas from Alberta, vertebrates underwent their second great evolutionary transformation. Fishes possessing jaws made from modified gill arches appeared.

Jawed fishes preyed upon less-well-equipped relatives and on each other.

Compared to their earlier diversity, few kinds of jawless fishes can be found living today. Modern survivors include naked-skinned lampreys and hagfishes.

JAWLESS DIVERSITY

Mark V. H. Wilson

Mark V. H. Wilson

Fork-tailed fishes such as *Furcacauda* and *Sphenonectris*, from the Northwest Territories' Mackenzie Mountains, differed in body shape, and possibly in swimming ability and lifestyle, from previously known jawless fishes.

Few exposures of vertebrate-fossil-bearing Palaeozoic rock exist in Alberta. The rocks recording that time were worn away long ago—along with the fossils they may have contained. Most of what we know about early fishes comes from sites in northern Canada, British Columbia, and the United States. The Mackenzie Mountains in the Northwest Territories are particularly rich in fossils of jawless fishes—adding to our understanding of the diversity of life in the seas of ancient western Canada.

Unlike previously known torpedo-shaped jawless fishes, fossils from the Mackenzie Mountains reveal new kinds of fishes, including some with heads that are small and conical, bodies that are narrow and deep, and tails that are deeply forked and up to half the length of the bodies. These fork-tailed fishes belong to a group known as thelodonts, jawless fishes whose bodies were covered with tiny, hollow scales.

"They actually would have resembled angelfish," says Mark Wilson, one of the University of Alberta palaeontologists who studied the unusual fossils. "We can't be certain, of course, but we think this body plan means they lived higher in the water column than other jawless fishes with which we're familiar. With that huge tail, they would have been powerful swimmers—slow, but powerful. And probably graceful." Large tails possibly helped the fishes avoid predators by improving swimming strength and speed, or they may have helped the animals power their ways through the muddy surface layers of the seafloor as they foraged for food.

The fork-tails are also the first fossil jawless fishes to reveal distinct stomachlike chambers and intestines within their guts. Before they died, the fork-tails enjoyed last meals of muddy seafloor sediments that later fossilized and preserved the shape of the gut. Wilson speculates that the stomachlike chamber indicates a dine-and-dash feeding strategy. The fishes may have descended to the seafloor to gobble food and store it in their guts, and then gone elsewhere to digest it.

GIANTS OF THE ORDOVICIAN UNDERWATER

About 445 million years ago, the western underwater margin of what is now North America was home to giants—giant sponges. Up to five metres long, the creatures towered over all other known organisms of the time. These animals belong to an extinct group of sponges known as stromatoporoids, which secreted a brittle, calcium-carbonate skeleton. Balancing on their ends and filtering tiny organisms from seawater, the sponges lived at the edges of the continent near the long slopes into the deep basin of the western ocean. Because the sponges lacked roots to anchor themselves to the seafloor, waves caused by storms or earth tremors occasionally knocked the animals down as if they were bowling pins, to be buried by mud.

The specimens now displayed at the Royal Tyrrell Museum have travelled a great distance since they lived. The collision between North America and the crust lining the Pacific Ocean 140 million to 80 million years ago drained the western edge of the continent and pushed the ancient seafloor towards the sky. Museum palaeontologist Paul Johnston found these fossils among the peaks of British Columbia's Top of the World Provincial Park, about 60 kilometres west of the Alberta border.

Paul A. Johnston, Royal Tyrrell Museum

Paul A. Johnston, Royal Tyrrell Museum

Palaeontologist Dr. Paul Johnston reclines beside one of the fossil sponges in eastern British Columbia's Top of the World Provincial Park, providing scale to the giant specimens (bottom). The fossils are of stromatoporoids (top), sponges that excreted calcium-carbonate skeletons.

WHERE IN ALBERTA

Athabasca Falls, south of the town of Jasper, displays Alberta's past from the early Palaeozoic Era. Here, the thundering cascade has cut through and exposed rock layers that make up the Gog Group. The tonnes of sediment that were washed from North America's Precambrian-aged granite hills and carried to the ocean by lifeless rivers during the early Palaeozoic are preserved in the red and pink quartzites of the canyon

Mike Todor

walls. Made almost entirely of quartz crystals, quartzite is one of the hardest rocks found in the Rocky Mountains. In some places, the Gog Group preserves spectacular burrows and trails of animals such as trilobites and feathery sea pens.

One of Alberta's most spectacular views of remnants of the early Palaeozoic is along the Trans-Canada Highway south of Lake Louise. Colours and weathering patterns on Castle Mountain's cliffs and slopes reflect the different environments of Early and Middle Cambrian seas. The thick, vertical cliffs of the Cathedral and Eldon formations are limy-shoal deposits. The beiges and pinks of the cliffs indicate where limestone was steam-cooked into dolomite before the rise of the Rocky Mountains. The sloping grey shales between the cliffs are ancient muds, clays, and silts that washed down from coastal lowlands during brief retreats of the sea. The fossil-bearing Stephen Formation is the dark, sloping layer between the Cathedral and Eldon formations.

Wil Andruschak

The quartzite found at Jasper National Park's Athabasca Falls (top) was formed by sediment washed into Alberta's Cambrian seas from the province's older granite hills. Farther south, along the Trans-Canada Highway, Castle Mountain (bottom) provides spectacular views of other preserved sea environments of the Cambrian Period.

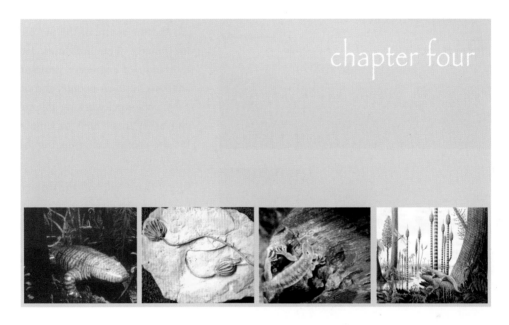

chapter four

LAND IS INVADED: THE LATE PALAEOZOIC

We owe our wealth, technology, and standard of living to billions upon billions of algae, bacteria, and other tiny marine creatures that lived and died in the seas that covered Alberta hundreds of millions of years ago. These creatures not only fuelled early ecosystems by feeding fish and invertebrates but have, in the intervening time, become the fuel that makes modern society possible.

ANCIENT SEAS, MODERN WEALTH: THE DEVONIAN

Tim Schowalter

Dennis Braman, Royal Tyrrell Museum

Dennis Braman, Royal Tyrrell Museum

The gleaming glass-and-concrete towers rise towards the sky from the banks of the Bow River. This is the centre of the city of Calgary . . . and the centre of the petroleum industry in Canada. The buildings are filled with thousands of engineers, geologists, computer analysts, business people, and other oil-company personnel working to provide oil, gas, and other fossil fuels to supply our hunger for the energy that runs our cars and heats our homes.

More than half of the mineral resources mined or extracted in Canada come from beneath Alberta. The province's resources include salt, cement, lime, sand, and gravel, as well as some metals such as gold. Most of the industry in Alberta, however, focuses on fossil fuels—oil, gas, coal, and their byproducts. The province's energy resources account for trillions of dollars in trade, royalties, taxes, and sales every year.

We owe our wealth, technology, and standard of living to billions upon billions of algae, bacteria, and other tiny marine creatures that lived and died in the seas that covered Alberta hundreds of millions of years ago. These creatures not only fuelled early ecosystems by feeding fish and invertebrates but have, in the intervening time, become the fuel that makes modern society possible.

We call the liquid remains of these lowly organisms petroleum.

Petroleum extraction (top) and refinement (middle) is a major Alberta industry. Millions of years ago, however, what we call petroleum was sea-scum. The city of Calgary (bottom) is considered the hub of the petroleum industry in Canada.

Royal Tyrrell Museum

LIFE AMONG THE REEFS

Towering 30 metres above the dark seafloor mud, a long, white ridge reaches for the surface. Sponges, sea lilies, and corals live in the sunlit waters, anchored safely on the reef away from currents and waves. Clams and worms burrow in the soft, limy mud that collect in depressions. Looking for tasty invertebrates, primitive jawed fish and squidlike ammonoids probe at the structure while trying to avoid sharks and armoured fish. The water teems with plankton.

Strange, limestone-encased creatures live across the top and down the slopes of the reef. Bulbous-bodied and bizarre-looking, they are stromatoporoids, the engineers and architects of this underwater highrise ecosystem. In areas where tides cause rough, surging currents, robust varieties thrive. Quiet, protected areas are home to slighter, more fragile kin. Others yet prefer deeper, dimmer waters, while shallow water is host to a different kind.

Wherever on the reef they make their homes, it is the lime minerals secreted by their bodies that add to the structure, building it higher towards the light. Their home stretches and branches for hundreds of kilometres. Like many other giant reefs that grow on floor of the great ocean bay that covers much of Alberta, it is an underwater island of life and tropical brightness in a sea of dark, cool water and suffocating muds.

Almost 380 million years later, the fossilized remains of these sunlit communities are the natural storage tanks for the hydrocarbons that fuel our economies.

Like modern reefs, Alberta's Palaeozoic structures provided shelter and food for entire communities of algae and animals. Sea lilies, trilobites, corals, ancient squids, and sharks all made their homes among these sunlit ecosystems that stretched across southern Alberta.

SOFT SKELETONS

Royal Tyrrell Museum

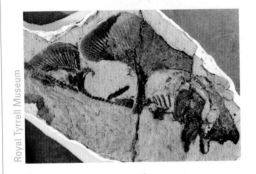

Royal Tyrrell Museum

Royal Tyrrell Museum

The streamlined, sea-dwelling predators that we call sharks first appeared during the Devonian Period, the Age of Fishes. They made their homes among the great stromatoporoid reefs. During the last 390 million years, their modern descendants retained many of the same features of early sharks. Sharks, skates, and rays lack both swim bladders, to adjust buoyancy, and gill covers. Their skeletons are made of cartilage.

Cartilage is a tough, flexible body material that rarely fossilizes. Fossils of ancient sharks and rays are limited almost entirely to teeth, which were shed constantly, and tiny barbed scales that covered the animals' bodies. Only extraordinary conditions at the time of a shark's death permit preservation of the skeleton.

Seventy-five million years ago, in what was to become Dinosaur Provincial Park, freshwater skate *Myledaphus* encountered perfect fossilization conditions. The barbs embedded in the fish's skin are in place, preserving the outline of the animal. Its jaws are batteries of six-sided teeth—probably used to crush molluscs in ancient waters. The backbone is articulated. The technician who prepared the specimen needed a high-powered microscope to see the specimen's details while he freed it from the surrounding sandstone. He even had to break grains of sand apart before removing them from the specimen's skin.

The skeletons of sharks, rays, and skates are made of cartilage, the same flexible material that is found in human noses and ears. Only exceptional conditions at the time of death permit cartilage fossilization, as happened with this shark (top) and with *Myledaphus* (middle). Tiny barbs once embedded in the *Myledaphus's* skin fossilized in place and now show the outline of the ray fish's body. Batteries of teeth (bottom) are also preserved.

RECIPE FOR PETROLEUM

- In a warm, shallow sea, mix billions of microscopic marine organisms.

- Set aside for several tens of millions of years while the ingredients live, reproduce, and die. When thick beds of organic ooze layer seafloor sediments, gently stir in bacteria that feed on organic remains.

- Let sit until most of the oxygen and nitrogen is removed.

- Slowly heat the mixture to between 75° and 125° Celsius by applying pressure in the form of kilometres of rock layers and sediments piled on top. Be careful not to overheat the mixture by burying it too deeply or undercook it by applying too little pressure, as either will result in failure.

- As the mixture cooks, remove it from the shale it formed in by squeezing it into more porous rock. Devonian reef rock, made from the remains of stromatoporoids, is preferable, although any rocks with spaces between sediment grains will do.

- Cover with cap rock, such as a salt dome or impermeable shale, until ready to extract.

The holes in this piece of reef rock are called vugs. They occur in the fossilized remains of reef-building stromatoporoids and were filled with white dolomite crystals after the reefs were buried. Where reef rock from the Devonian Period lies two kilometres beneath Alberta's prairie, vuggy dolomite is frequently saturated with petroleum, creating oil and gas reservoirs.

Royal Tyrrell Museum

Alberta baked in the sun at the start of the Devonian Period, enjoying a 10-million-year break between ocean submersions. Being dry land meant that many of the province's sedimentary rocks that had formed in earlier times were worn away and washed westwards to the coast that lay beyond Alberta's borders. In many places in the province today, Devonian-aged rock rests upon rock laid down during the Cambrian Period—bracketing a gap of 150 million years.

Evolving Earth: Late Palaeozoic

Pangaea means "one world." The Permian Period saw the planet's continents merge into one giant landmass that scientists now call Pangaea. The meeting of the continents assisted the spread of species on land and in the surrounding sea.

The ocean returned once again, and Alberta's rock record started reaccumulating. This time, however, the edge of the North American continent had been lifted by collisions with Greenland, Europe, and Siberia. In the west, a ridge stretching from Montana to Peace River, called the West Alberta Arch, contained the sea to a long inlet that crept from the north. Punctuated by retreats lasting hundreds of thousands of years, the ocean flooded valleys and basins with seawater. During retreats, lakes and lagoons larger than Utah's Great Salt Lake evaporated under Alberta's hot, tropical sun. Deposits of salt, potash, and gypsum formed.

North America continued its slow, spinning drift to the north, and the West Alberta Arch gradually eroded and disappeared. Another collision, this time between western North America and the crustal plate under the Pacific Ocean, thrust up a chain of volcanic islands far offshore which resembled the islands of Japan or Indonesia. As the ocean plate ground its way beneath North America, a trench formed between islands and mainland.

Life on Land

Royal Tyrrell Museum

By the end of the Devonian Period, forests of tree-sized horsetails and ferns housed and fed spiders, millipedes, and springtails. It was onto these swampy landscapes that lobe-finned fishes first ventured from the water.

From tiny *Cooksonia*-like stock evolved plants with complex features: big, flat leaves that improved photosynthesis, roots that aided absorption of nutrients and anchored plants to the ground, woody tissue that increased size and stability, and seeds that protected new generations of vegetation. Plants were joined by insects whose ancestors left the sea for new opportunities on land—including shelter, and greenery on which to feed. Fossils of Devonian insects preserved within stems of fossil plants suggest the two groups of land-dwelling organisms influenced each other's evolution.

The stage was set for vertebrates to leave the water and walk on land. In the ponds, lakes, and rivers of the northern Devonian continents, a group of bony fishes was evolving. These fishes could

Europe and Asia merged, creating the Urals. With that, Laurasia was born, forming one of two landmasses that dominated Earth during the late Palaeozoic Era. The southern continents also regrouped. Africa, Antarctica, India, Australia, and South America joined to become Gondwanaland.

By the end of the Palaeozoic Era, Alberta was part of Earth's single, giant landmass known as Pangaea. As continents ground into one another, more mountain ranges were raised towards the sky. Shallow seas drained. Climates and environments became more distinct from region to region: ice sheets covered the Southern Hemisphere until well into the Permian Period, deserts claimed equatorial regions, and swamps and forests blanketed middle and high latitudes. Conifers and tree ferns became abundant in cooler, drier regions of the planet.

Alberta was oceanside property again, looking northwest across the Panthalassa, the ocean that covered 70 percent of the Earth's surface.

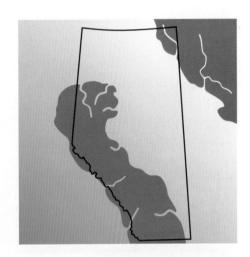

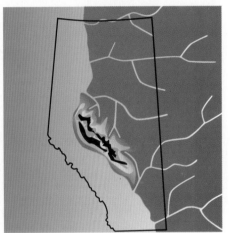

During the Late Devonian (top) the West Alberta Arch rose 300 metres above sea level and forced seas to flood Alberta from the north.

Throughout the Late Carboniferous and the Permian periods (bottom), the coastline moved back and forth across western Alberta, creating gaps in the geological record of Permian-aged rock

breathe through both gills and lungs, and they had leglike lobes to power their fins. Of all of these unique fishes, the rhipidistians likely were the first to emerge onto land. They lived in shallow water, they could breathe air if their water became too muddy, and they used their fleshy fins both to stabilize themselves in fast-moving currents and to negotiate muddy riverbanks.

From there, it was just a few awkward steps onto land.

Royal Tyrrell Museum

Coelacanths were once candidates for the position of ancestor to the first amphibians. They have limb-like, lobed fins, just as their close cousins, the rhipidistians, do. Coelacanths were long thought to have become extinct during the Cretaceous Period, about 70 million years ago. In 1938, a live specimen was caught off the coast of Africa—the first of more than a dozen modern coelacanths captured. The fossil specimen shown here dates from about 220 million years ago.

MAKING TRACKS IN THE ARCTIC

Royal Tyrrell Museum

Royal Tyrrell Museum

In 1994, scientists searched Ellesmere Island's Strathcona Fiord for traces of animals that lived 375 million years ago (top). Scorpion footprints and tail-drag marks and millipede trackways from the High Arctic island now adorn the walls of the Royal Tyrrell Museum's Palaeozoic Era exhibit (bottom).

The sun does not set in the Arctic at the height of summer. From the beginning of June to the end of July, Ellesmere Island's midnight landscape glows in the light of a sun that rests low on the horizon.

The island has many outcrops of rocks that formed about 375 million years ago, during the Devonian Period. Royal Tyrrell Museum Curator of Invertebrates Paul Johnston spent three weeks on the island during the summer of 1994, prospecting and collecting fossils. Every day, he and his colleagues hiked between camp and quarry, following a trail they had made through the rocks and tundra. Every day, they skirted the edges of the same boulders and stepped over the same slabs of rock.

One morning, Johnston arose earlier than usual and set off towards the quarry. Because of the hour, the sun was lower in the sky. The slightest relief in the terrain made shadows. The light bathed a siltstone slab lying next to the trail. Johnston glanced at it as he passed, looked again, and stopped. Shadows of tiny bumps and pits patterned the surface of the grey rock. They were the footprints and tail-drag marks of ancient invertebrates.

The fossil slab and several others from the same area are now displayed at the Tyrrell Museum. Although Johnston cannot confirm what creatures made the tracks, he thinks they may have been millipedes.

WHERE IN ALBERTA

Remnants of a Devonian community loom over the Icefields Parkway, Highway 93, in Banff National Park. The Weeping Wall, one of Cirrus Mountain's sheer cliffs, is one of the most accessible and most spectacular exposures of the Devonian-aged Fairholme Group. Between waterfalls that cascade down the cliff, the ancient limestone wall reveals skeletal remains of fossilized corals, algae, stromatoporoids, brachiopods, and other creatures.

Some of Alberta's most photographed mountain views include exposures of the Palliser Formation. The impressive cliff of Jasper National Park's Roche Miette and the lower cliffs of Rundle and Three Sisters mountains along the Trans-Canada Highway east of Banff are dark-coloured limestone deposited during the Late Devonian, after Alberta's stromatoporoid reefs had disappeared. Jasper's Maligne Canyon cuts through Palliser rock.

There are many places among Alberta's mountains to view rocks and fossils laid down during the Devonian Period, including Jasper National Park's Weeping Wall (top) and Maligne Canyon (bottom).

Dennis Braman, Royal Tyrrell Museum

Alberta Community Development Parks & Protected Areas, Kananaskis Country

WHERE IN ALBERTA

Other reef fossils are seen in road cuts overlooking the uppermost Grassi Lake, above the town of Canmore. The cliff located in the gap between Rundle Mountain and Ha Ling Peak is studded with stromatoporoid fossils.

One of Alberta's few examples of preserved land environments from the late Palaeozoic can be found in the Yahatinda Formation on the east face of Wapiti Mountain near Sundre. These ancient deposits consist of sandstone, siltstone, and limestone conglomerates set down along river channels that flowed west across the West Alberta Arch into the Panthalassa Ocean. Geologists have collected broken fragments of jawless fish and reed-like plants.

Alberta Community Development, Parks & Protected Areas, Kananaskis Country

Dennis Braman, Royal Tyrrell Museum

In Kananaskis Country near Canmore, Ha Ling Peak (top) towers over some of southern Alberta's most accessible fossil beds that date from the Devonian Period.

Plant and fish fossils can be found on the slopes of Wapiti Mountain (bottom) near Sundre.

OUT OF THE SEA, ONTO LAND: THE CARBONIFEROUS

Royal Tyrrell Museum

Plants' new, efficient leaf designs, branching patterns, roots, and seeds paved the way for the development of the complex communities characteristic of the Carboniferous Period. Forests of towering, 30-metre-tall club mosses, horsetails, ferns, and seed-bearing trees sheltered understoreys of ferns and other smaller plants. Swamps covered much of Europe and eastern North America. These provided the plant material that produced the massive coal beds for which the Carboniferous Period is named.

No longer dependent on water for fertilization, seed-bearing plants invaded drier landscapes, aided by wind and insects that dispersed seeds. As plants became better at surviving on land, they transformed hostile environments into places that provided food and shelter for insects and early land-dwelling vertebrates.

But no matter how inviting plants made dry landscapes, they could not entice amphibians far from water. Throughout their 350-million-year history, amphibians have remained dependent on the ponds, rivers, and streams where they hatched. Even today, amphibians must lay their shell-less eggs in water or other moist environments to prevent them from drying out. The eggs hatch into tadpoles that breathe through gills and swim like tiny fishes. As they mature, most amphibians lose gills and fins, grow legs, and develop lungs.

The swamps that thrived in the swampy lowlands of coastal North America and Europe during the Carboniferous and Permian periods were perfect habitat for amphibians. There they ruled, some growing as large as modern crocodiles.

Swamps of the Carboniferous Period (bottom) were home to giant dragonflies, huge scorpions and spiders, metre-long millipedes and cockroaches, and other invertebrates. These creatures provided food for many new vertebrate species evolving on land, including *Ichthyostega* (top), one of the earliest known amphibians.

Royal Tyrrell Museum

Royal Tyrrell Museum

The earliest known reptile remains are fossilized in Carboniferous-aged tree trunks in Nova Scotia. Early reptiles, called cotylosaurs, were small and compact, and they looked much like modern lizards.

Other animals evolved that severed their ties to rivers and ponds. By the end of the Carboniferous, about 290 million years ago, the first vertebrates able to live entirely on dry land had emerged. These were reptiles, cold-blooded animals with dry, scaly, waterproof skin. Protective coverings around their eggs freed them from the tyranny of water. A hard or leathery eggshell prevents damage and dehydration, and the amniotic sac within allows a reptile embryo to develop in a miniature, portable copy of its ancestral river or pond. These innovations allowed late Palaeozoic reptiles to colonize new territories: high, dry landscapes and the deserts of the Laurasian interior.

SMALL THINGS

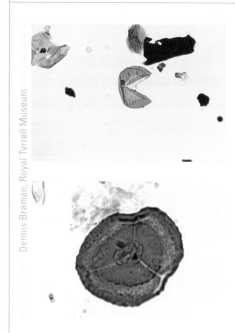

Dennis Braman, Royal Tyrrell Museum

Plants in the same family not only look similar when they are grown but their pollen or spores retain common family features. Shape and size are passed down through generations, evolving over time as new plant species evolve.

Much of our knowledge of ancient plant evolution is based on palynological evidence, the record of microscopic fossils such as spores and pollen.

Plants produce spores and pollen to reproduce. These tiny capsules contain blueprints of parent plants' genetic information. Because the information carried by pollen and spores is vital for a plant species to continue, plants have evolved ways to wrap the reproductive messengers in protein armour. As a result, the capsules are almost indestructible in oxygen-poor environments such as rivers, streams, and swamps. Time, water, and the pressure of millions of tonnes of rock do not harm them.

A single small sample of sedimentary rock can harbour thousands of these tiny fossils.

By analyzing the palynological record, scientists such as Royal Tyrrell Museum palynologist Dennis Braman know that ferns appeared about 350 million years ago, about the same time as horsetails and other kinds of scouring rushes. Cone-bearing plants—conifers—emerged shortly after and became dominant forest plants during the Permian, when climates became drier. They remained plant kings of the forest until the Cretaceous Period, when flowering plants emerged.

WHERE IN ALBERTA

The easily eroded Banff Formation and the craggy Rundle Group are Alberta's most familiar record of Carboniferous times. These rocks make up the jagged peaks of many of south-western Alberta's mountains. Grotto and Pigeon mountains and Mounts Lougheed, Norquay, and Rundle can be seen along the Trans-Canada Highway east of Banff. For views of the spectacular Opal Range, travel Highway 40 through Kananaskis Country. Horn coral fossils are common at Ptarmigan Cirque in Peter Lougheed Provincial Park. In the Livingstone Range between Crowsnest and Highwood passes, fossils of ancient sea lilies are found in Banff shales and Rundle limestones. The scattered remains look like bits of stone cereal and pretzels embedded in rock. Complete fossils are occasionally found; these resemble small mops. Although sea lilies evolved during the Cambrian Period, they became especially common in the shallow seas bordering Alberta during the Early Carboniferous.

Dennis Braman, Royal Tyrrell Museum

Alberta Community Development, Parks & Protected Areas, Kananaskis Country

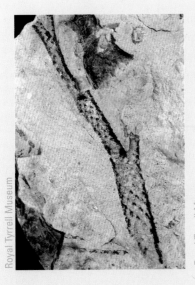

Royal Tyrrell Museum

Royal Tyrrell Museum

Jagged peaks such as Mount Rundle (top) and the Opal Range (middle) contain remnants of Alberta's Carboniferous Period. Fossils of 325-million-year-old horned corals, brachiopods, and sea lilies (bottom right) can be found in the mountains' shales and limestones. Farther north, Carboniferous shales entomb pieces of swamp-loving *Lepidodendron*, a scaly-trunked tree which grew up to 30 metres in height (bottom left).

ONE LAND, ONE SEA: PANGAEA AND THE PANTHALASSA DURING THE PERMIAN

Rise of Reptiles

Royal Tyrrell Museum.

Small, swamp-dwelling cotylosaurs were the stock from which all known reptiles evolved, including the ancestors of turtles, lizards, and ichthyosaurs. During the Permian Period, one group came to rule the reptile world: the mammal-like reptiles. Big, sprawling pelycosaurs such as sail-backed *Dimetrodon* appeared about 280 million years ago and hunted smaller animals for about 30 million years.

Therapsids, a more diverse group of mammal-like reptiles, began their reign in the Permian. By the end of the period, they were the dominant dryland herbivores and carnivores. Some may have been warm-blooded and grown coverings of fur or hair over their bodies. Others had evolved jaws and teeth similar to those of many modern mammals.

While *Dimetrodon* and its mammal-like kin ranged across Pangaea's vast territories, a group of small, agile, narrow-jawed reptiles sought a better life in water. *Mesosaurus* took to lakes and rivers to prey on fish and conquer new, vacant territories. With all major landmasses welded together into supercontinent Pangaea, other reptiles soon followed *Mesosaurus*, spreading quickly across the continents.

Dimetrodon, with its bladelike teeth, is a member of the group that gave rise to mammals. The reptile likely used the sail-like ridge on its back to control body temperature in northern Pangaea's hot climate.

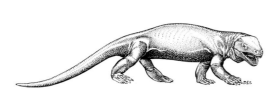

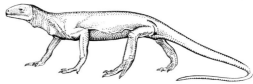

As reptiles evolved, some species marked the evolutionary transition of amphibians into reptiles. *Seymouria* (above) is considered an amphibian, but it has skeletal traits characteristic of reptiles.

Many modern vertebrate groups emerged in the Palaeozoic. *Petrolacosaurus* (above) began the long lineage of animals that includes archosaurs, lizards, dinosaurs, and birds.

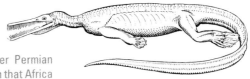

Fossils of *Mesosaurus* (right) in freshwater Permian sediments in South Africa and Brazil confirm that Africa and South America were joined 260 million years ago.

A Bad Ending

Two-hundred-fifty-million years ago, everything suddenly changed. Ninety percent of all species in Earth's seas vanished. Marine animals such as corals, sea lilies, brachiopods, and squidlike ammonoids were especially hard hit by the extinction. The long reign of trilobites ended; armoured fish did not survive.

On land, 70 percent of animal species disappeared. The toll includes three-quarters of reptiles and six of nine major groups of amphibians.

Fossil pollen and spores indicate that plant life also dramatically changed. Plants that produced seeds lacking protective coverings disappeared almost entirely. A sharp increase in fossil spores suggests fungi were treated to a feast of dead organic material immediately after the extinction. In layers of rock laid down shortly after, new dry-climate pollen suddenly appears.

The extinction is so noticeable in the fossil record that it is used to mark the boundary between two great eras of life on Earth: the Palaeozoic (Ancient Life) Era and the Mesozoic (Middle Life) Era. Scientists have long been mystified about what caused such an abrupt change. Theories include chemical imbalance in the ocean due to poor circulation, or climate change caused by dropping sea level and increased volcanic activity.

Recently, however, new evidence has been uncovered: the mass extinction marking the end of the Palaeozoic and the beginning of the Mesozoic occurred quickly—within one million years. And, according to molecular research on sediments laid down at the time of the extinction, a visitor from outer space—an asteroid or a comet that struck the planet—may have caused it. Cagelike carbon molecules in the sediments have been found to encase 250-million-year-old gas atoms: the chemical signatures of the atoms match those found only in meteorites and comets.

Little other evidence for extraterrestrial impact exists from this time. Scientists supporting the theory believe that may be because the comet slammed into the Panthalassa, the ocean that surrounded Pangaea. An ocean impact would have prevented most signs of collision from forming, and the impact site itself would have been recycled into Earth's mantle long ago.

IT CAME FROM OUTER SPACE

Alan Hildebrand

Tiny crystals of shocked quartz provide evidence that a giant asteroid struck the coast of Mexico 65 million years ago. The grain shown here measures one-third of a millimetre across.

In the mid-1980s, geologists identified a 65-million-year-old meteor crater centred on the small village of Chicxulub, Mexico. The crater measures 80 kilometres across and about 20 kilometres deep. More significantly, however, the timing and location of the crater's creation corresponds to evidence of sudden disaster in sediments around the world.

It also corresponds to the disappearance of dinosaurs from the fossil record.

Here was hard, physical evidence that the extinctions marking the end of the Cretaceous Period 65 million years ago were a result of a sudden, global catastrophe, not slowly changing climates or shifting continents.

For decades prior to the discovery, geologists had been collecting evidence supporting the impact theory. Shocked quartz, microdiamonds, and tiny glass droplets form in high-pressure, high-heat, explosive events, such as when an asteroid or comet strikes the Earth's crust. These were found in a broad arc of end-Cretaceous sediments stretching northwest across the North American continent. The mineral iridium is rare on Earth. Its abundant presence in extinction-aged sediments around the world was additional evidence for a major impact. Sharp increases in fern spores in the fossil record dating from shortly after time of impact indicate that huge amounts of vegetation died suddenly and provided opportunities for some plants to quickly colonize newly vacant landscapes.

The asteroid that slammed into Mexico 65 million years ago would have had dramatic, worldwide effects on climate and life—the sea boiling at the impact site, firestorms caused by burning fallout raging across North America, dust catapulted into the atmosphere blocking sunlight for months . . .

The crater's discovery prompted many scientists to take closer looks at other ancient global extinctions. Although much more evidence is needed, impact markers in fossil sediments may implicate extraterrestrial impacts in at least five mass-extinction events. Traces of iridium, shocked quartz, and glass spherules in China and western Europe coincide with a worldwide extinction of many coral-reef communities and other plants and animals in the Late Devonian. Extraterrestrial chemical signatures of atoms trapped in sediments laid down at the end of the Permian Period match the timing of the most severe extinction faced by life on Earth. Fossilized marine-invertebrate diversity drops suddenly at the end of the Triassic—corresponding to a time when a series of impact craters were created in eastern and northern Canada. Some scientists suspect that a series of impacts in the eastern United States, Siberia, and across Europe may have contributed to a mass extinction that occurred about 35 million years ago.

Not enough evidence has been gathered to prove impacts played roles in any of these extinctions. Even for the most-studied impact–extinction event—the one ending the Age of Dinosaurs—research shows that diversity of plants, dinosaurs, and other animals was steadily lessening for several million years prior to impact. For many species, the asteroid that struck ancient Mexico may have been only the trigger that shifted an already-teetering balancing act between survival and extinction.

©2002 Museums and Collections Services, University of Alberta

Alan Hildebrand

Meteorites frequently strike the planet's surface. Most are no bigger than dust particles, but occasional larger examples, such as these that fell near Mayerthorpe in 1964 (top) and Bruderheim in 1960 (middle), survive the journey through the atmosphere. Asteroid impacts are much rarer and can cause greater damage. About 50,000 years ago, a small iron asteroid struck Arizona's Colorado Plateau, forming a crater about one kilometre across (bottom).

WHERE IN ALBERTA

Alberta's record of the Permian Period is scarce. That time in the province's history can be traced along the Bow River near the Banff Springs Hotel. Here, ancient rocks document the changing coastline of ancient Alberta. Gaps in the geological record where rocks were exposed and worn away interrupt layers of shallow-water rock.

The low cliffs below Bow Falls and on the east bank of the river present one of Alberta's few exposures of phosphate-rich rock. Phosphate is thought to form from shallow-water accumulations of crustacean shells and fish teeth, bones, and scales rich in phosphorous. The Bow Falls phosphate layer is thin, dense, and dark. It is part of the Johnston Canyon Formation, which also forms the cliffs of Johnston Canyon on Highway 1A between Banff and Lake Louise.

Dennis Braman, Royal Tyrrell Museum

The cliffs below Banff National Park's Bow Falls offer glimpses of environments that existed in Alberta during the Permian Period.

REPTILES ASCENDANT: THE EARLY MESOZOIC

Early in the Triassic, four groups of land-dwelling reptiles took to the sea. Nothosaurs, thalattosaurs, ichthyosaurs, and placodonts thrived in their new environment. Within 20 million years, primitive ichthyosaurs with webbed feet and eel-like bodies were replaced by ichthyosaurs with bodies similar to those of dolphins, with paddles for limbs, and with teeth and jaws perfected for eating fish and squids. They lived near the coast and among chains of islands.

MESSENGER FROM THE PAST

The Rolex Awards for Enterprise/Photographer Tomas Bertelsen

Palaeontologist Betsy Nicholls reaches across the skull to examine an eye socket or a ridge of bone. The skull alone is as big as a table. It weighs 1.5 tonnes. Ten people could dine comfortably around its crushed surface. The animal's vertebrae are large enough to serve as dinner plates. Placed end to end, as in life, they would stretch across the floor of the cavernous lab and out its doors.

The beast that these bones belong to is a behemoth of the blue ocean. Measuring 23 metres in length—slightly smaller than a blue whale—it is the world's largest known ichthyosaur, a reptile that swam the globe-girdling ocean called the Panthalassa during the Mesozoic Era, 225 million years ago. It lived by hunting fish and squidlike animals called belemnites, its dolphin-shaped body moving through the water, its upright, forked tail helping it chase quick-swimming quarry.

Today the predator is imprisoned in sulphurous-smelling shale and silt-stone from the banks of the Sikanni Chief River, near British Columbia's Pink Mountain. Technicians from the Royal Tyrrell Museum are liberating the beast from its rocky bonds. The animal's bones and the rock that encases them contain clues to a watery world long gone, to a group of extinct reptiles that started off on land and turned to the sea, and to ancient, underwater ecosystems that existed off the coast of Alberta during the Triassic Period, 250 million to 205 million years ago.

Palaeontologist Dr. Betsy Nicholls (top) examines the world's largest known ichthyosaur fossil (illustrated at right) in the Royal Tyrrell Museum lab while technician Wendy Sloboda removes rock from the fossil. More than 20 high-power-helicopter trips were required to transport the remains of the 23-metre-long marine reptile specimen from the remote excavation site in northeastern British Columbia.

A TRYING TRIASSIC TIME

Bound by two mass extinctions and witness to the breakup of supercontinent Pangaea, the Triassic Period was a time of beginnings and endings. The global extinctions that ushered in the 185-million-year Mesozoic Era—the era of Middle Life—left holes in Alberta's ocean ecosystems and, to a lesser extent, in land environments.

Survivors moved in to fill gaps. Many were animals that had filled marginal roles or lived in harsher habitats within pre-extinction communities. With previously dominant species removed or reduced in numbers, hardier organisms had an opportunity to colonize new areas. New behaviours and new body forms emerged as survivors accessed resources previously unavailable to them. Over time, some behaviours and forms proved better than others at exploiting new niches, and new species evolved.

This is how ecosystems recover from major extinctions. Experimentation is usually followed by a period of stabilization during which species sort themselves out and settle into niches. Inevitably, some species are better suited to surviving and reproducing within the environment than others: these outcompete less-successful species, initiating a second, smaller extinction event. Eventually, long-term stability occurs: successful lineages evolve slowly, few new lineages arise, and few older ones become extinct.

According to research by scientists at the Geological Survey of Canada, the recovery of shallow-water ecosystems in Early Triassic Alberta is documented by traces left by invertebrates. In rocks that formed soon after the Permian extinction, certain kinds of brachiopods and their trace fossils are abundant. As the rocks get younger, brachiopod trace fossils are gradually replaced by trace fossils of diverse groups of creatures that included bivalves, worms, and sea lilies.

Albertonia's (top) elaborate pectoral fins may have powered short bursts of speed, stirred up food from ocean-floor muds, or even helped to attract mates.

During medieval times, people thought belemnites' rod-shaped remains were devils' fingernails. Palaeontologists have found a more mundane identification for the pencil-shaped skeletons (middle): they are the remains of an ancient relative of squids.

Ammonoids (bottom) diversified so quickly throughout the Mesozoic that they serve as markers of biological time for the era. By identifying the species of ammonoid preserved within sedimentary rocks, scientists can pinpoint when the rocks formed to within about one million years—an eye blink compared to Earth's entire history.

SURVIVORS OF THE GREAT EXTINCTION

Tim Schowalter

While scientists have a fairly good understanding of the Triassic Period in Europe, little is known about what was going on during that period in North America—particularly in northwestern North America. In fact, the Early Triassic is one of the last geological intervals to be studied in detail in western Canada.

In the 1990s, geologists began systematically studying Alberta's Triassic outcrops—found mainly in the Rocky Mountains. For several summers, John-Paul Zonneveld of the Geological Survey of Canada travelled the 3,000-kilometre length of the Rocky Mountain Front Ranges, collecting and studying samples of rocks that had formed during the Early Triassic. Although the rocks, as expected, contained few fossilized animals—only ammonoids, clams, and the teeth of tiny animals called conodonts—Zonneveld found a wealth of trace fossils. The burrows, tracks, fossilized dung, and other signs of behaviours of invertebrates that lived in the seas that covered the western part of Alberta from about 210 million to 240 million years ago help to provide a more complete picture of life in the Triassic of western North America.

Some of the traces look as if they may have been made by trilobites, which were long thought to have become extinct at the end of the Permian. Others suggest behaviours of lobsterlike decapods, worms, and ancient pillbug-like animals called isopods. Fossils suggesting activity by numerous kinds of bivalves and brachiopods are common.

As time passed, the sea off the coast of Alberta became home to the animals that characterize the advancing Triassic Period. Ray-finned fishes such as *Albertonia* and diamond-shaped *Bobasatrania* joined sharks, coelacanths, and other, more-familiar-looking fishes. Coiled ammonoids and straight belemnites, both relatives of modern squids, added to the variety of sea life.

Long, slender nothosaurs were fish-eating marine reptiles with long necks and webbed feet. They became extinct at the end of the Triassic Period.

Although Zonneveld is often unable to identify exactly what animals made the traces, he says the fossils provide valuable information on how animals adapted to environments that had been changed by the mass extinction at the end of the Permian. "The traces were made by the survivors and their descendants," he says. "They show great diversification in not only territories but in behaviours: both entire species and individuals within species were using new ways to access resources and take advantage of new opportunities."

Alberta's outcrops are so rich in fossil material that scientists from the Geological Survey, the Royal Tyrrell Museum, and numerous universities are conducting a detailed, multiyear study of Triassic-aged rocks throughout western Canada, rock layer by rock layer.

"The more we study these rocks," says Zonneveld, "the more we realize that a great many more animals actually survived the Permian extinction than we had thought, only to be wiped out in the extinction at the end of the Triassic."

Geologist John-Paul Zonneveld (right) collected samples of Early Triassic Period rocks from the length of the Rocky Mountain Front Ranges (opposite page, top). The rocks contained trace fossils (right, inset) of invertebrate animals that survived and diversified after the extinction that wiped out ninety percent of sea creatures at the end of the Permian Period.

The most notable additions to the Panthalassa's seascapes were marine reptiles. Early in the Triassic, four groups of land-dwelling reptiles took to the sea. Nothosaurs, thalattosaurs, ichthyosaurs, and placodonts thrived in their new environment. Within 20 million years, primitive ichthyosaurs with webbed feet and eel-like bodies were replaced by ichthyosaurs with bodies similar to those of dolphins, with paddles for limbs, and with teeth and jaws perfected for eating fish and squids. They lived near the coast and among chains of islands.

Thalattosaurus probably spent most of its time in coastal waters, hunting for small fish and sea jellies to eat, and occasionally emerged to bask on rocks and beaches, as marine iguanas do today.

TOMBS IN THE SKY

Royal Tyrrell Museum

Ganoid Ridge, looming high over Wapiti and Fossil Fish lakes, is a seafloor tomb come to rest amidst the clouds. Its black shales, siltstones, and fine-grained sandstones contain remains of fishes, seagoing reptiles, and marine invertebrates that lived in the waters of the Panthalassa off Alberta's coast during the Early Triassic, 205 million to 250 million years ago.

Scientists from the Royal Tyrrell Museum and the University of Alberta embarked on several expeditions to the eastern British Columbia region in the mid-1980s. They uncovered and collected an uncrushed coelacanth skull and more than 20 kinds of fishes. Many had never before been identified and are now believed to be evolutionary links between the ancient fishes of the Palaeozoic and modern ray-finned fishes.

They also found marine reptile fossils. Ichthyosaur skeletons from Wapiti Lake help Betsy Nicholls, marine vertebrate palaeontologist at the Royal Tyrrell Museum, answer questions about the evolution of this group of sea reptiles. A specimen of *Chauhusaurus* found in Triassic rocks above Wapiti Lake includes a leg and foot which are midway to becoming a true ichthyosaur paddle. In the few million years between *Chauhusaurus* and slightly more recent ichthyosaurs, some foot bones were lost, new bones were added, and other bones were fused together to form a distinct paddle. The speed of this transformation emphasizes how quickly the rate of evolutionary change was occurring among ichthyosaurs once the reptiles took to the oceans. Within 20 million years, they had assumed the streamlined, dolphin-shaped proportions of advanced ichthyosaurs.

The Wapiti Lake site proves many of the animals that lived there during the Triassic were both more diverse and more widely distributed than had been thought. Of the ichthyosaurs, *Chauhusaurus* is previously known from China, and *Utatsusaurus* is known also from specimens found in Japan. Thalattosaur fossils are found in northern California and Switzerland, as well as in northeastern British Columbia. The fish fossils found near Ganoid Ridge are similar to those found in Triassic sediments of the same age in many countries on different continents: Africa's Madagascar, the arctic Spitsbergen, and Greenland. At the time the rocks were being deposited, these places were located on the edges of supercontinent Pangaea.

Chauhusaurus (right, top) marks the transition within a group of animals between a time when ancestors of ichthyosaurs lived both on land and in the ocean, and a time when ichthyosaurs were fully marine animals—living, sleeping, hunting, breeding, and giving birth to live young in the ocean.

All of the Wapiti Lake thalattosaurs (right, bottom) are preserved in sandstone, suggesting that Ganoid Ridge (above) was covered by a shallow coastal sea 225 million years ago.

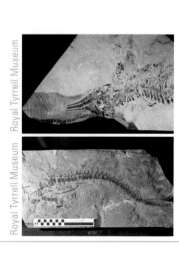

Royal Tyrrell Museum

Royal Tyrrell Museum

PINK MOUNTAIN

In 1992, Fort St. John archaeologist Keary Walde reported finding large fossil bones in the banks of the Sikanni Chief River, across the British Columbia border from Manning, Alberta. When marine vertebrate palaeontologist Betsy Nicholls checked the site, she found that the bones belonged to an ichthyosaur. Fortunately, little of the specimen was exposed or in danger of eroding away quickly, allowing Nicholls several years to work out the difficulties of excavating and collecting a large specimen from the remote northern location.

Nicholls returned to the banks of the Sikanni Chief River with a field crew in 1999, with backing from the Tyrrell Museum, Japan's National Science Museum, and the Discovery Channel. The river was in flood, threatening to sweep crew and bones from the rock shelf that contained the specimen; in the cold, wet weather, air compressors, jack hammers, and rock saws failed—too far from the nearest town to be easily fixed. Despite the difficulties, by the end of the first year's expedition, Nicholls and crew had excavated the fossil's 1.5-tonne skull and several of its massive vertebrae.

They returned again in following summers to collect the rest of the animal. It took more than 20 helicopter trips to carry the fossils to the nearest road, where they were loaded on trucks and driven to the Tyrrell Museum in Drumheller.

"It looks like it may be a new kind of ichthyosaur," says Nicholls. "We won't know for certain until the whole thing is prepared and studied, but the size suggests it."

Large ichthyosaurs are known from California and Nevada, but those specimens are only about 15 metres long. The Sikanni Chief River animal measures 23 metres long—almost as large as a blue whale.

Photographer Tomas Bertelsen

The Rolex Awards for Enterprise •

Palaeontologist Dr. Betsy Nicholls (top) and crew braved bad weather, failing equipment, and the flooding Sikanni Chief River (bottom) during the expedition's first summer collecting the ichthyosaur fossil.

Tim Schowalter

The Mesozoic Era—the era of Middle Life— is divided into three periods of geological time: the Triassic Period, the Jurassic Period, and the Cretaceous Period. Together they represent 185 million years of Earth's history. During the Mesozoic, continents assumed positions almost recognizable today, and ancestors of many of the animals who share the planet with us—crocodiles, birds, turtles, mammals, and frogs (top), for instance—made their first appearances.

Triassic Landscapes

Much of Pangaea's interior was arid. Along the coast, part of which bisected Alberta, parched seasons alternated with wet, monsoonlike extremes. The climate encouraged new kinds of plants to evolve and fill gaps left on land by the extinctions at the end of the Permian Period. Dry-climate plants such as cycads, ginkgoes, and cone-bearing conifers spread quickly. Seed ferns and horsetails added to the vegetation.

New kinds of plants meant new kinds of animals to feed on them and shelter within them. The Triassic Period saw a quick succession of reptile dynasties overtake one another. Remnant populations of great mammal-like therapsids survived the Permian extinctions to make a comeback. However, they were never again as numerous nor as diverse: as tens of millions of years passed, they were replaced by archosaurs, which included crocodile-like animals called thecodonts, dinosaurs, and pterosaurs.

Early dinosaurs were small and are believed to have walked on their hind legs. This would have freed their forelimbs to help to capture food or for support when feeding. Dinosaurs quickly evolved into two distinct groups, each of which had a different arrangement of bones within the hips. Lizard-hipped dinosaurs included both predatory herrerosaurs and long-necked, plant-eating prosauropods. Spiky scelidosaurs and small, fleet-footed fabrosaurs numbered among the Triassic's relatively rare, plant-eating, bird-hipped dinosaurs.

Mammals emerged at about the time the first dinosaurs appeared. Tiny, shrewlike ancestors of humans scurried about in undergrowth and trees, feeding on insects and seeds, and keeping wary eyes open and noses twitching for hungry reptiles.

Sikannisuchus huskyi is one of very few existing specimens of animals that may have lived in Alberta's landscapes during the Triassic Period. An early relative of both dinosaurs and crocodiles, it has the body of a crocodile and the large, serrated teeth of a meat-eating dinosaur. Scientists cannot say for certain how the animal lived: it may have dwelt on Alberta's ancient beaches and swam through waters in estuaries along the province's coast, 220 million years ago, or it may have spent all of its time in the ocean.

Making Way for Terrible Lizards

The Triassic Period was no time to get comfortable. By 225 million years ago, the continents that made up Pangaea began to separate. The resulting changes in geography and local climates may have tipped the balance of survival for many species.

A series of asteroid impacts across the northern continents about 220 million years ago may have triggered the mass extinction near the end of the period. Fossils of marine invertebrates found on British Columbia's Queen Charlotte Islands indicate the extinctions happened in a heartbeat of geological time—in less than 50,000 years. Snails, clams, and ammonites suffered great losses, while nothosaurs, thalattosaurs, and placodonts disappeared completely. About half of the major groups of land-dwelling reptiles, including most thecodonts, died out.

Ancestors of modern crocodiles, turtles, and frogs survived. Small and inconspicuous, mammals also continued. Dinosaurs suddenly were able to capitalize on their superior, bipedal mobility by becoming the top predators and the top planteaters in the Mesozoic world.

NASA

The formation of the impressive Manicouagan crater in Quebec dates from near the end of the Triassic Period. At 70 kilometres across, it is ringed by an enormous lake.

Shocked quartz found in three different layers of marine rock laid down in Italy at the end of the Triassic Period indicate at least three asteroids bombarded Earth. They may have been part of one giant asteroid that was ripped apart by the planet's gravity before entering the atmosphere, as Comet Shoemaker-Levy had in 1994. More evidence, however, is needed to prove that the end-Triassic mass extinctions were caused by asteroids.

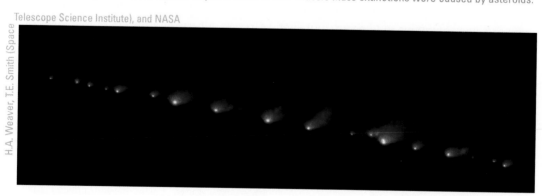

H.A. Weaver, T.E. Smith (Space Telescope Science Institute), and NASA

For the first 50 million years of the Mesozoic Era, Alberta's dry, eastern lowlands sloped gently down to the Panthalassa Ocean, which covered the western part of the province. Broad, sweeping rivers crossed the province's flat, muddy eastern plains, picking up mud, clay, and sand, and releasing them in coastal waters. The rocks that formed from these deposits document frequent changes in water level and temperatures: deeper, cooler waters alternated with warm, shallow-water environments and hot, dry tidal flats as shorelines retreated into British Columbia.

Evolving Earth: Early Mesozoic

Great monsoon storms blowing off the oceans hammered coastal Alberta during the Triassic Period. Beginning about 210 million years ago, the continents forming Pangaea began to pull apart, creating the Atlantic Ocean.

Although Pangaea reached from pole to pole and across the equator, Alberta rested north of the tropics, on the supercontinent's western shore. Long dry spells and annual monsoons marked the seasons. With Pangaea's entire landmass at its back and the ocean to the west, Alberta bore the brunt of seasonal storms blowing onto the coast.

By the end of the Triassic Period, 205 million years ago, cracks were showing in the supercontinent. Great rift valleys thinned the crust between Europe and North America, and between North America and Africa. Where land ripped open, deep canyons formed. The Panthalassa flowed into the valleys, becoming the early Atlantic Ocean.

North America drifted northwestward. Pushed by the growing Atlantic Ocean, it travelled about 1,000 kilometres in 30 million years. By mid-Jurassic, Alberta had reached its present latitude—far from the heat of the tropics it had known for most of its existence. Sea levels rose, and southern Alberta, Saskatchewan, and Manitoba were flooded under a new, warm, shallow seaway.

In its journey across the surface of the planet, North America encountered other landmasses. Beginning in the Jurassic and continuing until about 70 million years ago, Alberta and the entire west coast of

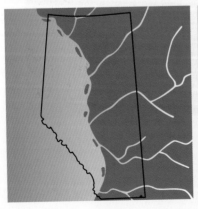
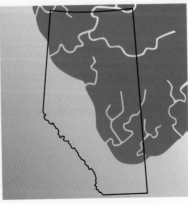

Alberta's (left) coastline during the Triassic Period stretched from where Hay River is in the north to where Lethbridge is in the south.

During the Jurassic Period, the Pacific Ocean drained from Alberta (right). Forests grew on the resulting swampy plains, providing raw materials for coal formation.

North America began a series of slow head-on collisions with at least five chains of volcanic islands anchored on the spreading floor of the Pacific Ocean. Like a bulldozer driving into oncoming traffic, Alberta rammed the islands and scraped them off the ocean-floor pavement. The province's underwater margins were lifted out of the water, broken, and bent. Where fractures in the continent reached deep into the Earth, rivers of magma flowed towards the surface, cooking surrounding rocks and turning into underground oceans of granite. Pressure and heat sutured the island chains to the continent's leading edge. North America acquired most of British Columbia, California, Oregon, and Washington in this way.

As massive sheets of the continent's edge jackknifed skywards and thrust eastwards, they rode up over one another as if they were three-kilometre-thick roofing shingles. Their combined weight caused the floor of Alberta to sag. This westward-dipping Alberta Basin stretched from northern British Columbia to Montana. During breaks between collisions, erosion of the new highlands outpaced the rate of mountain building. As the land to the west wore away, its sand, mud, and silt washed to the eastern basin and western ocean, and the weight of the rock sheets eased. Alberta's basement rebounded—slowly—raising the sediments towards the sky.

Then, along came another island chain . . .

Gradually choked with Jurassic sediments, the Alberta Basin saw the final, 20-million-year draining of the Pacific Ocean from the continent. Forests grew on swampy plains in the warm, humid climate. As the basin sank under the growing weight of rising mountains and sediment, the swamps were buried.

The Alberta Basin remained a prominent feature of the province's landscape and geology until the end of the Tertiary Period, two million years ago.

By the end of the Jurassic, Pangaea was no more. Earth was a watery world dotted with familiar-looking island continents. The Tethys, the giant seaway that separated northern and southern landmasses, circulated equator-warmed waters around the globe, bringing warmth and rainfall to mid-latitude regions such as Alberta. Even at the North Pole, which then laid in northern Canada, forests covered the land. Here temperatures were mild throughout the year, but seasonal differences in the amount of sunlight, as can be seen in polar regions today, affected photosynthesizing plants. Species of ginkgoes and conifers, which dominate the region's Late Jurassic plant-fossil record, adapted to long polar-winter nights by evolving the ability to shed and regrow leaves each year.

The weight of mountains building on the continent's edge during the Jurassic Period caused sagging farther inland. During the 140 million years that followed, the resulting basin filled with sediment.

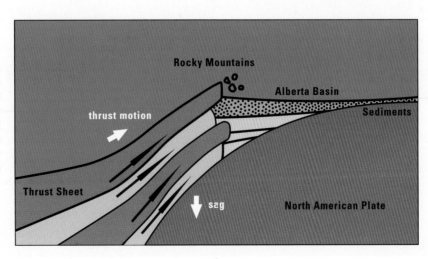

WHERE IN ALBERTA

The Triassic Period is represented in Alberta rocks by the Spray River Group. Ranging in colour from inky grey and bright red to beige-white, the siltstone, sandstone, limestone, and dolomite are remnants of a time when shallow, muddy seas advanced and retreated over the province's flat coastal plains.

Dark, sheeting Triassic siltstone found in the southern Rockies is prized building material; the Banff Springs Hotel, the Banff park administration building, and parts of the Banff Avenue bridge are faced with it.

To see the siltstone in its natural rock layers, visit Bow Falls below the Banff Springs Hotel. Thrusting from the west bank of the falls, sheets of the sharply angled rocks are partially buried under their own eroded flakes and chips. The rocks are soft and are easily worn away by the force of the water. The siltstone can also be seen in outcrops along the Spray River, south of the town of Banff.

Highway 11 near Kootenay Plains follows the surface traces of the Sulphur Mountain Thrust Fault. The red rocks that vertically band the summits of the mountains looming over the highway are Sulphur Mountain Formation siltstone. These same rust-coloured rocks are seen tilted almost vertically in road cuts along Highway 40 near the junction to Peter Lougheed Provincial Park. The ripple-marked rocks are remains of a shallow seafloor that existed off the coast of Alberta more than 220 million years ago.

Wil Andruschak

Andrew Neuman, Royal Tyrrell Museum

Wil Andruschak

Rocks dating from the Triassic Period can be seen in the walls of the Banff Springs Hotel (top), along the Spray River Gorge (middle) south of Banff townsite, and near Kootenay Plains (bottom).

JURASSIC INNOVATIONS
Dinosaurs at Home around the World

The Jurassic Period witnessed many new developments in life. Although few terrestrial rocks of this age remain in Alberta, fossils from other parts of the world show that the extinction of many groups of land-dwelling animals at the end of the Triassic Period created new opportunities for survivors, just as had happened at the beginning of the Triassic. Dinosaurs emerged as the victors of the later extinction and quickly assumed recognizable forms during the Jurassic: long-necked sauropods, killing-machine theropods, and fleet-footed, vegetarian camptosaurs.

Variations in dinosaur body shapes and lifestyles were encouraged by the growing isolation of the planet's landmasses. During the Triassic, animals had had free range of continents. With Africa's Ivory Coast sutured to Florida and Georgia, and the journey from Nova Scotia to Spain only footsteps long, there had been few barriers to the range of animals and plants across Pangaea. Fossils of similar Mesozoic dinosaurs entombed in different parts of the world today archive this past proximity—fossils of prosauropod plateosaurs are found in Canada, Europe, Greenland, and China, and remains of small, carnivorous Syntarsus are found on both sides of the Atlantic.

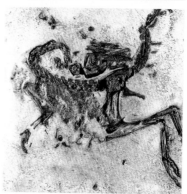

This changed during the Jurassic. Over millions of years, as landmasses broke away from Pangaea and inched their ways across the globe, each continent carried its share of the mix of life over the surface of the planet to far-flung regions. In isolation on the island landmasses, groups of organisms charted their own evolutionary paths as they responded to changing environments on their home continents.

The rivers, deltas, coastal forests, and swamps that replaced the dry inland landscapes of the Triassic meant shelter and food for new breeds of dinosaurs.

Despite the variety in shapes, sizes, and behaviours, all dinosaurs retained common physical traits that define them as a group. When identifying a Mesozoic-aged reptile fossil, palaeontologists look at how hips, ankles, wrists, and jaws are put together. Very specific arrangements, shapes, and sizes of bones in these parts of a skeleton are the characteristics that distinguish dinosaurs and birds from every other group of animals.

Meat-eating terrors such as 10-metre-long *Allosaurus* (top) and tiny *Ornitholestes* (middle-top) made their debut during the Jurassic Period. Some of the largest animals to ever walk on land left footprints across southern North America: long-necked sauropods such as *Camarosaurus* (middle-bottom) weighed up to 90 tonnes each—the equivalent of almost 20 school buses. Stegosaurs (bottom) warmed themselves with heat-absorbing plates lining their backbones.

Royal Tyrrell Museum

The skeletal traits dictated how dinosaurs moved and lived. The structure of bones in the hips and ankles meant dinosaurs moved with their legs directly under their bodies, unlike lizards and crocodiles, which hold their elbows and knees out from their sides. It also meant dinosaurs walked on their toes, as can still be seen in their modern descendants, birds. Dinosaurs were land-dwelling creatures: except for birds, they did not fly, as pterosaurs did, nor did they live in oceans, as ichthyosaurs and plesiosaurs did.

In addition to that, all dinosaurs, again except for modern birds, lived only during the Mesozoic Era.

Arrangement of bones in the ankles, hips, and feet meant that dinosaurs' legs supported the animals' hips from below (left). In other reptiles, legs are held out to the sides, giving the animals a sprawling gait (right)

Dinosaurs Take Flight

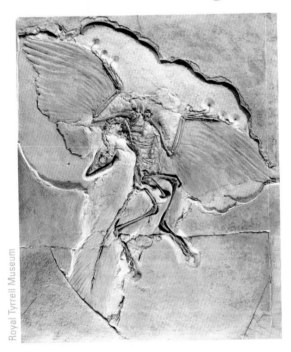

Archaeopteryx was the first fossil bird to be discovered. Half-bird, half-dinosaur, it combines features found in birds and meat-eating dinosaurs, and it sparked debate about the evolutionary relationship between the two groups of animals.

One day in 1996, while studying fossils in Beijing, Royal Tyrrell Museum dinosaur palaeontologist Philip Currie was taken aside by Chinese colleagues. His hosts wanted to show him the fossil of a new kind of dinosaur that recently had been found in the northeastern corner of the country.

Currie opened the silk-covered box presented to him. Inside was a small meat-eating dinosaur. It had teeth, claws, and arms similar to those of *Troodon* or *Dromaeosaurus*. Along the back of the head, neck, and back of this specimen, however, were imprints of what appeared to be downy feathers.

"I was astounded," says Currie. "I'd always expected feathered-dinosaur fossils to exist somewhere: it was just a matter of time before one was found. But, that day in Beijing, it was the last thing I expected."

The little feathered dinosaur was *Sinosauropteryx*. It was the first feathered-dinosaur fossil to be found in northern China. Following that discovery, a dozen other, similar fossils were found in the same region. Each strengthens the theory that modern birds are closely related to dinosaurs—a theory first proposed in 1870 when *Archaeopteryx* was discovered in limestone deposits at Solnhofen, Germany. The Jurassic bird's teeth, long bony tail, clawed fingers, and unmistakable impressions of flight feathers suggested a creature that was half-dinosaur, half-bird. Many scientists at the time believed *Archaeopteryx* was evidence of the evolution of dinosaurs into birds.

Found alongside *Archaeopteryx* in the Solnhofen deposits were fossils of ammonites, twigs, pterosaurs, and the small meat-eating dinosaur *Compsognathus*. Re-examination of the animal fossils a century later revealed that two of the specimens had been misidentified. They, too, were fossils of *Archaeopteryx*. The skeletons of the two animals—one, a bird; the other, a dinosaur—are so similar that scientists had confused them.

Since then, analyses of bird skeletons and theropod fossils show the two groups of animals have more than 120 skeletal traits in common with each other—and with no other animals. Physical characteristics shared exclusively among different species indicate common ancestry: the more numerous the common characteristics, the closer the evolutionary relationship. In birds and small theropods, the shape, size, and arrangement of dozens of bones in the backbone, ankles, hips, shoulder, hands, and feet are almost identical. The bones of both animals are hollow and thin-walled. Both animals have wishbones and both lay similar eggs. The Chinese discoveries prove that some dinosaurs had feathers, as birds do.

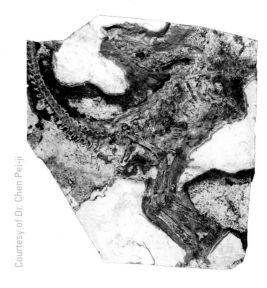

Courtesy of Dr. Chen Pei-ji

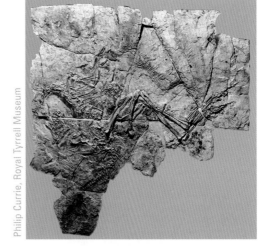

Philip Currie, Royal Tyrrell Museum

Sinosauropteryx (top) is the first-known specimen of a feathered-dinosaur fossil.

Specimens of *Protarchaeopteryx* and *Caudipteryx* (middle) preserve feathers that may have helped the animals attract mates. Chinese scientists have also discovered fossils of *Confuciusornis* (bottom), a bird with flight feathers, and of *Sinornithosaurus*, the first-known feathered dromaeosaur.

Philip Currie, Royal Tyrrell Museum

AERIAL COMPETITION

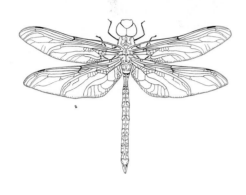

Birds were not the first animals to take to the air. For hundreds of millions of years, insects reigned supreme as creatures of powered flight.

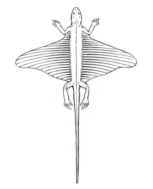

During the Permian Period, a small group of reptiles called coelurosauravids developed wings from extended ribs covered by skin. These reptiles used their wings to glide through the air. *Wapitisaurus problematicus*, the fossil found by Tyrrell Museum staff at Wapiti Lake in the 1980s, may be a gliding reptile.

Pterosaurs were the first backboned animals to use powered flight, beginning about 210 million years ago, during the Triassic. A pterosaur wing consists of an extremely long fourth finger that was used as a strut. This supported a sheet of leathery skin that stretched from fingertip to hip. Another membrane of skin from hand to neck provided additional lift and stability. *Quetzalcoatlus* had an 11-metre wingspan, making it the largest known flying creature.

Birds appeared during the Jurassic, and are thought to be closely related to small meat-eating dinosaurs. Unlike the kite-shaped pterosaur wing, a bird's wing consists of a short arm, a fused hand, skin, and feathers. Feathers are modified scales. They may have first developed for display and to help dinosaurs to control body temperature. Only later would they have helped in the evolution of bird flight.

Swamp Litter and Coal Dust

For more than a century, coal has played influenced the lives of people residing in southwestern Alberta's Crowsnest valley. In 1878, George Dawson, a geologist surveying a route through the Rocky Mountains for a new railway, reported an abundant coal seam. By the end of that century, coal mining fed families in more than a dozen communities along the valley. Every day, long trains of railcars filled to the brim with Crowsnest coal were pulling out of the valley to feed the country's hunger for fuel.

Things have changed in the last 50 years: tourism has overtaken mining as the region's main industry. However, plenty of coal remains in the hills. The richness of the seams archive the lush forests and swamps that existed in the area long ago. Southwestern Alberta owes more than half a century of prosperity to beds of partially rotted vegetation that collected and were buried in swamps and bogs from 165 million to 100 million years ago.

At that time, spanning the end of the Jurassic and the beginning of the Cretaceous periods, British Columbia was slowly building on the western edge of North America as island chain after island chain rammed the continent and thrust the continent's edge eastward towards Alberta in a series of long mountain ranges. The floor of western Alberta sank under the weight of the mountains. At times, deltas formed at the mouths of huge rivers that meandered from eastern Alberta and from the rising western highlands. Forests grew thick: conifers, cycads, tree ferns, and ginkgoes sheltered ferns and shrubs. As swamps, the forests would have flooded seasonally, encouraging the accumulation of thick beds of leaf litter and dead trees.

During the course of almost 65 million years, as mountains rose in fits and starts to the west, the land repeatedly sagged and rebounded. Thick beds of rotting forest vegetation accumulated and were buried under hundreds of metres of mud and silt carried in by the rivers. With time and the accumulation of sediments above, the swamp litter was transformed into coal.

Tim Schowalter

Alberta's Crowsnest Valley (top) contains seams of coal that were once swampy forest of conifers, cycads, and tree ferns.

NA–3903–136, Glenbow Archives

Working with heavy equipment in constricted spaces, bad air, and rock dust took their toll on the men who mined Jurassic Period coal (bottom) from mountains surrounding the Crowsnest valley. For more than 50 years in the nineteenth and twentieth centuries, the federal Department of Mines and Minerals issued statistics showing the number of tons of coal extracted from the region for each fatality or serious injury.

BIG FEET IN ALBERTA

Mining along the Alberta–British Columbia border has uncovered more than rich coal deposits. The coal beds have also yielded the fossil trackways of dinosaurs and other extinct animals.

Along the Alberta–British Columbia border, some of the richest coal seams that stripe the rock west of the Crowsnest Pass contain mines of information about ancient animals that lived near the boundary between the Jurassic and Cretaceous periods. Sandstone and mudstone layered among the coals record the long-ago presence of mid-Mesozoic animal communities. At one particular site, about 130 million years ago, small reptiles tracked across soft ground and swam through muddy shallows. Birds and pterosaurs strutted. Meat-eating dinosaurs visited the area. Giant long-necked sauropods with feet up to one metre across paced.

Long after these animals departed the region—long after they died and their species became extinct—their tracks remained, footprints imprinted in the rock as if they were shadows on photographic film.

Today, natural sandstone casts and moulds of the prints are being recovered and studied by Richard McCrea of the University of Alberta's Ichnology Research Group, which studies trace fossils made by ancient animals.

Trace fossils such as footprints and trackways record an animal's presence in an area. Bones and shells may be scattered by scavengers or washed away from where the animal died, but footprints—similar to fingerprints at a crime scene—are an animal's personal testimony of having been in a particular place at a particular moment.

"It was always thought sauropods never made it this far north," McCrea says. "There are very few northern occurrences elsewhere. Absolutely no skeletal material of sauropods has ever been found anywhere in Canada, and yet footprints tell us these animals lived here after all."

This fossil is a natural cast of a footprint made by a long-necked sauropod about 130 million years ago in the Crowsnest region. It measures almost one metre across, and about 20 centimetres deep on one side of the footprint.

Some scientists believe the geographic range of different plant-eating dinosaur groups was split north–south. From the middle of the Cretaceous Period until its end, hadrosaurs and ceratopsians dominated northern regions. In the south, sauropods ruled supreme; there was no evidence that they had ever ventured into what is now Canada. Climate and habitat may have limited where sauropods lived, although at this point, evidence is uncertain.

"Research out of the United States and the rest of the world says sauropods lived in drier climates in lower latitudes," says McCrea. "But here we have proof they frequented low-lying, well-vegetated, low-energy environments—coal swamps. This is one of a handful of sites in the world where sauropod prints are found in this sort of environment: sauropod habitats are not as straightforward as we had thought."

Although palaeontologists cannot identify species from tracks, the scientists can usually figure what kind of animal it was. Footprints are moulds of the bottoms of an animal's feet. Meat-eating dinosaurs, for instance, left three-toed tracks with sharp claw impressions. Footprint size indicates whether the theropod was large or small. Age and location of the sediments that preserved the tracks may narrow identification further—large, three-toed prints in rock laid down in Late Cretaceous Alberta were probably made by *Albertosaurus*, *Daspletosaurus*, or *Tyrannosaurus*.

When Royal Tyrrell Museum palaeontologist Philip Currie mapped and collected trackways from northeastern British Columbia's Peace River Canyon before the river was dammed in the 1970s, he learned the difference between tracks made by birds and tracks made by small meat-eating dinosaurs such as *Troodon* or *Sinornitholestes*. "In the dinosaur feet," says Currie, "the angle formed by the outer and inner toes is less than 90 degrees. In bird feet, it is usually greater than 90 degrees because of changes in bird feet for perching in trees."

Trackways also record behaviour. They tell us how an animal moved, in what direction, and whether it was alone or in a group with others of its kind. Sometimes trackways even tell us how fast an animal moved.

Before the Peace River Canyon was dammed and flooded in the 1970s, palaeontologists collected fossil trackways of birds and meat-eating dinosaurs from beside the river (left).

A trackways site near Grande Cache sloped steeply down into a coal mine, its precarious position a product of the formation of the Rocky Mountains 80 million years ago. Across the cliff, bird tracks crisscross trackways made by duckbilled dinosaurs, herds of ankylosaurs, and large and small theropods (right).

WHERE IN ALBERTA

Beneath much of Alberta's prairie, rocks that formed 100 million years ago during the Early Cretaceous lie directly atop Devonian-aged limestone deposited more than 250 million years earlier. Many of the rocks that had formed in these areas between the Devonian and Cretaceous periods were stripped away by erosion during the Mesozoic Era. Now these pages in the archives of Alberta's early Mesozoic past can be seen only in the Rocky Mountains and foothills.

The alternating beds of orange siltstone and dark shale thrusting skywards in road cuts along the Trans-Canada Highway east of the Banff exit and along Highway 40 north of Kananaskis Village catalogue the erosion of the highlands growing to the west of the Alberta Basin during the Jurassic. These beds are part of the Fernie Formation, which formed from sediments washed into the Alberta Basin 208 to 144 million years ago. Fossils of snails, clams, and ammonites can be found where these rocks are exposed along Ribbon Creek in Kananaskis Country. Farther south in the Crowsnest region, fossils of squidlike belemnites and ammonites have been found, including an ammonite shell almost one metre across.

Outcrops of the Fernie Formation can be seen in road cuts along the Trans-Canada Highway, east of the Banff exit (top). The cliffs of Kananaskis Country's popular Ribbon Creek (middle) contain fossil ammonites (bottom) from the Jurassic Period.

Wil Andruschak

Alberta Community Development, Parks & Protected Areas, Kananaskis Country

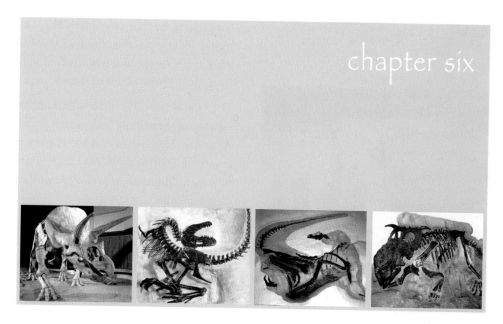

DINOSAURS COME OF AGE: THE LATE MESOZOIC

The most obvious record of changes in animal life in Cretaceous Alberta occurred among dinosaurs. A landbridge connecting northern Canada to Asia allowed an exchange of dinosaur groups. Protoceratopsians, whose Cretaceous fossils are common in the sands of modern Mongolian deserts, may have originated in North America, where they quickly diversified into horned dinosaurs such as *Centrosaurus*, *Styracosaurus*, and *Pachyrhinosaurus*. Ostrich-like ornithomimids also trekked over the top of the world. Ankylosaurs are found in Asia. Small theropods such as dromaeosaurs are found on both continents, while fossils of plant-eating duckbills can be found on almost every continent.

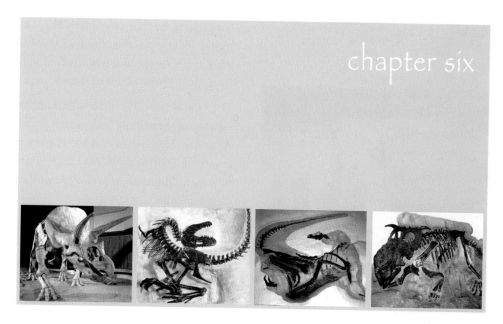

75

THE GREAT CANADIAN DINOSAUR RUSH

NA-3596-138, Glenbow Archives

American Museum of Natural History Library

Neg./Trans. No. 18546,

Royal Tyrrell Museum

Discoveries of dinosaur bones by Barnum Brown (top), a bone collector with the American Museum of Natural History, sparked the Great Canadian Dinosaur Rush in 1911. Brown (middle) and the Sternbergs (bottom) collected more than 200 specimens during their five summers in Alberta's badlands.

Staring at the cliff, the man surveys the sandstone, mudstone, and coal that climb 100 metres above his head to the prairie. Crowning the layer cake of rock is a nine-metre-thick icing of yellow clay. He notes the petrified wood and ironstone littering the water-worn base of the slope. He is particularly interested in the fossils eroding from the hill. Together, more than 40 bones form the tail of a duckbilled dinosaur, a two-legged plant-eater that lived in the area 70 million years ago. Splinters and fragments of other bones scatter the slope below the specimen.

The year is 1909. The man is Barnum Brown. He is a curator of vertebrate palaeontology at New York's American Museum of Natural History. Since his discovery of Triceratops in 1897 and Tyrannosaurus rex in 1902 in Montana, he is also one of the most famous dinosaur collectors in the world.

Brown is on his way home from digging for fossils in Montana, but he has detoured north to Drumheller to scout for dinosaur bones at the invitation of John L. Wegner, a local landowner. Wegner had visited the New York museum earlier in the year and told staff there that bones similar to those of the museum's giant dinosaurs could be found all over his ranch in Alberta. Wegner had tried to interest scientists in eastern Canada with his news, but nobody had pursued the matter.

Brown, however, was interested. And now that he has seen the fossils on Wegner's property, he is intrigued; the rock outcrops in which the bones are found are said to extend about 150 kilometres along the Red Deer River. This is enough to convince him to return the following summer to prospect for dinosaur remains for his museum.

Just 13 years after thousands of fortune seekers had passed through Edmonton on the overland journey to the Klondike in search of gold, Brown's 1910 field expedition down Alberta's Red Deer River sparks the Great Canadian Dinosaur Rush. Although dinosaur fossils had been known since the 1870s to exist in the province, little attention had been paid to the resource. It takes Brown's interest and the success of his 1910 and 1911 expeditions, during which dozens of dinosaur specimens are shipped to New York, to prompt the Canadian government to take interest in the fossil riches. In a bid to keep some of Canada's dinosaur remains in Canada, the government hires another American, Charles Hazelius Sternberg, and his three sons to prospect for and collect Red Deer River fossils.

Even at the height of the Dinosaur Rush, only about a dozen people descend upon Alberta to search out and claim the remains of ancient vertebrates—far fewer than the hordes who took part in the Yukon Gold Rush. And, unlike the Klondike Gold Rush, the Great Canadian Dinosaur Rush is civilized: there is no lawlessness; there is no brawling among men mining the riches; and field work stops on Sundays, when Brown and the Sternbergs host local residents visiting the field camps. Sometimes the palaeontologists even visit each other to swap stories and share pots of camp tea.

Despite the few people who participate, the Great Canadian Dinosaur Rush puts the country on the dinosaur-fossil map. More than 200 complete or nearly complete specimens are collected during the five summers that Brown and the Sternbergs labour under the hot prairie sun. And after Brown leaves Alberta for the last time, the Sternbergs continue collecting the area's exquisitely preserved skeletons—some for Ottawa's Geological Survey and National Museum of Canada, some for other museums and research institutions. Because of their efforts, Alberta dinosaurs are shipped to Toronto, Chicago, Los Angeles, London, Berlin, Paris, Buenos Aires, and other cities around the world. Each specimen is a silent scientific ambassador, announcing to the world the province's wealth in fossils that date from the last 15 million years of the Cretaceous Period.

FACE TO FACE WITH DINOSAURS

In 1884, the Geological Survey of Canada assig Joseph Burr Tyrrell to survey mineral resource throughout southern Alberta. He was instructed to pay particular attention to coal deposits that might power the transcontinental railway being built by the Canadian Pacific Railway.

While examining coal seams in the Kneehill Creek valley, 10 kilometres west of Drumheller, Tyrrell came face to face with the skull of *Albertosaurus sarcophagus*.

Today, just kilometres away, the Royal Tyrrell Museum of Palaeontology commemorates th discovery.

No. 201735- A. Reproduced with permission of the Minister of Public Works and Government Services Canada, 2002, and Courtesy of Natural Resources Canada, Geological Survey of Canada

Although he later explored Canada's arctic barrenlands, took part in the Klondike Gold Rush, and made his fortune in Ontario's Kirkland Lake gold mines, Joseph Burr Tyrrell is best remembered in Alberta for his part in the discovery of the province's graveyard of Late Cretaceous dinosaurs.

CRETACEOUS COMMUNITIES
Powerful, Flower-Full Alberta

Kevin Ross Aulenback, Royal Tyrrell Museum

Of the many families of flowering plants that existed when dinosaurs disappeared 65 million years ago, 90 percent are still with us today.

photos (clockwise): *Cercidiphyllum* fossil, *Cercidiphyllum*, Monocot flowers, ginger flowers, fossil tuber related to ginger, monocot leaves

Flowering plants are commonplace today. They include shrubs and weeds as well as broadleaved trees such as willows, aspens, and birches, and they can be found in every modern Alberta ecosystem. One-hundred-twenty-million years ago, however, flowering plants were just beginning to evolve. It was only during the last half of the Cretaceous Period that flowering plants refined their ability to take advantage of plant-munching animals—as an aid to reproduction and dispersal—to a high evolutionary art.

Flowers contain a plant's reproductive organs. Colourful and scented, they also advertise nectars and pollen to insects, birds, and other animals. When these animals pillage blooms for food, they carry pollen to other flowers of the same species, unintentionally assisting in the plants' reproduction. Flower pollination encouraged pollinators to diversify: beetles and social insects such as bees and ants became abundant during the Cretaceous. Today, they number tens of thousands of species each.

Another reproductive refinement by flowering plants is the herbivore-proof casings that protect the plants' seeds and seed factories. Once they worked out a way to keep seeds from being cracked open or digested, some flowering plants devised new ways to disperse seeds. Some evolved fruits to entice animals to dine and then deposit seeds in readymade fertilizers some distances away. Other flowering-plant seeds developed hooks and barbs with which they hitchhike rides on the fur or feathers of passing animals. Long-distance seed dispersal gives flowering plants an edge over those that rely solely on wind or water to colonize new territories.

In addition to these strategies, flowering plants are both fast-growing and forgiving. Their seeds sprout faster and their seedlings grow faster than those of ferns, conifers, cycads, and ginkgoes. Flowering plants are also more tolerant of being trampled or chewed on by animals. Today, a field disturbed by overgrazing or wildfire grows flowering plants before any other plant recovers.

However, it took millions of years for environments around the world to separate flowering plants with successful strategies from those that couldn't compete. In the Early Cretaceous, flowering plants formed only a small part of the planet's vegetation. By the end of the period, they had changed the look of Alberta. A landscape once dotted with evergreens had been replaced by one covered in fast-growing brush and broadleaved trees mixed with conifers.

Many palaeontologists believe the growing importance of flowering plants in Cretaceous ecosystems went hand in hand with the evolution of new forms of plant-eating dinosaurs whose remains are so common in Alberta's Cretaceous-aged rocks. Duckbilled and horned dinosaurs began to be common during this time. Their diversity and numbers may have been influenced by the abundance and growing strategies of tender, tasty flowering plants

Alberta Dinosaurs

Animals thrived in the forests and open spaces of the plains that extended from Alberta's western border to the coast of the inland sea. Small reptiles such as crocodiles, champsosaurs, and turtles lived on riverbanks and in swamps. Lizards, salamanders, and social insects became established. Diving birds such as *Baptornis* fished along the coast, while other birds found homes farther inland alongside pterosaur *Quetzalcoatlus*. Marsupial and placental mammals crept about at dusk, searching for insects and seeds.

The most obvious record of changes in animal life in Cretaceous Alberta occurred among dinosaurs. A landbridge connecting northern Canada to Asia allowed an exchange of dinosaur groups. Protoceratopsians, whose Cretaceous fossils are common in the sands of modern Mongolian deserts, may have originated in North America, where they quickly diversified into horned dinosaurs such as *Centrosaurus*, *Styracosaurus*, and *Pachyrhinosaurus*. Ostrich-like ornithomimids also trekked over the top of the world. Ankylosaurs are found in Asia. Small theropods such as dromaeosaurs are found on both continents, while fossils of plant-eating duckbills can be found on almost every continent.

Each of these animals had its own ecological home in ancient Alberta's ecosystems. Some thrived in wetlands. Others preferred drier areas. Some may have migrated with the seasons, moving inland for wet, stormy seasons, and returning to the coastal plains for drier seasons.

Fossil evidence suggests that some dinosaurs developed complex social behaviours and lived in groups. Fleet-footed duckbilled

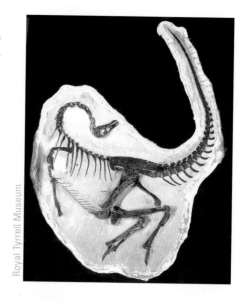

Royal Tyrrell Museum

In 1995, scientists discovered this specimen of bird-mimic dinosaur *Ornithomimus*. The skeleton is 98 percent complete, and it includes a complete skull with fossilized remnants of a beak. Study of this and other specimens tells scientists that ornithomimids were closely related to tyrannosaurs and troodontids.

As the pileup of Pacific islands that had started during the Jurassic Period continued along the western edge of North America, the sagging of Alberta's basement rocks reached across Alberta into Saskatchewan. The Alberta Basin sank deeper into the planet's mantle; erosion of the rising western highlands converted British Columbia's newly exposed ancient seafloors and volcanic rock into mud, sand, and silt that poured into the depression. In some parts of the basin, layers of sediments more than three kilo-

Evolving Earth: Late Mesozoic

During the Cretaceous Period, the growing Atlantic Ocean pushed North America towards the north-west, and the southern continents separated.

metres thick were deposited. Rivers flowed eastwards across this landscape, ending in giant coastal estuaries and deltas. During the early part of the Cretaceous, a polar sea much warmer than today's Arctic Ocean covered northern Alberta.

In British Columbia, volcanoes wormed their ways through the fractured crust and erupted, blowing clouds of dust and ash into the wind to settle across Alberta. Today, the ash is preserved as bentonite clay. It is frequently found in Alberta's badlands: when dry, it appears crumbly; when wet, it is sticky, butter-slick mud. Its abundance indicates an almost unbelievable amount of volcanic activity during the Cretaceous Period.

The separation of supercontinent Pangaea continued. As the Cretaceous began, South America was tearing away from Africa, and Antarctica and Australia were setting their own courses across the plan-

The Western Interior Seaway was a major feature of Alberta's landscape and weather patterns during the Cretaceous Period.

et's surface. North America continued its slow separation from Europe. Meanwhile, northwestern North America joined with northeastern Asia, allowing periodic exchanges of animals throughout the next 120 million years.

The processes deep within the planet's crust that had caused the continents to separate also displaced ocean water, causing seas to rise and spill across the land. By mid-Cretaceous, sea level measured almost 250 metres higher than it is today. The sagging Alberta Basin invited the northern ocean to creep south over what are now the prairie provinces. The water inched closer to seas invading the continent's interior from the south, finally meeting and bisecting the continent 110 million years ago. The joining of arctic and Gulf of Mexico waters created a 4,000-kilometre-long, 500- to 1,000-kilometre-wide seaway that brought new, warm-water marine species to Alberta.

Royal Tyrrell Museum

Volcanoes that erupted in British Columbia during the latter part of the Cretaceous Period provided the ash that today forms southern Alberta's bentonite clay. When wet, bentonite is as slippery as butter; when dry, it resembles popcorn.

Throughout the remainder of the Cretaceous, the seaway swelled, shrank, and shifted its coastline, making its presence felt across Alberta for 35 million years. Warm and shallow, it probably never exceeded depths of 300 metres. However, because of its size, it controlled the region's weather. Its warmth moderated the effects of Alberta's northern latitude, encouraging a moist, Florida-type climate on the coastal plain that extended from the sea's western shore. Cool air flowing down from British Columbia's infant mountains corkscrewed around columns of warm, humid air rising over the water—brewing storms and hurricanes. The weather nourished forests of ginkgoes, redwoods, cycads, and tree ferns as well as new kinds of flowering broadleaved trees such as magnolia, fig, sycamore, and primitive birch and maple.

Kevin Ross Aulenback, Royal Tyrrell Museum

Royal Tyrrell Museum

Modern ginkgo leaves lie against a southern Alberta ginkgo fossil (left). Another modern descendant of a tree that grew in Alberta during the Late Cretaceous Period is the China fir, shown (center) with fossil cone (right).

Royal Tyrrell Museum

Rapid burial in fine-grained sediments preserved the texture of skin on the underside of this duckbilled dinosaur 75 million years ago. This duckbill is known as *Corythosaurus* ("helmet reptile") for the helmet-like crest on the animal's head.

dinosaurs nested in inland colonies. Many duckbill species had crests atop their skulls or large chambers within their noses and crests which may have made communication by sound possible over long distances or within forest environments. Horned dinosaurs sported elaborate frills at the back of their skulls and spikes that may have helped the dinosaurs recognize each other, select mates, or defend territory. The jumbled collection of remains of groups of some kinds of dinosaurs indicates these species died in groups, and also may have lived in groups.

Meanwhile, behaviours of other dinosaurs remain mysteries.

Charting the Course of Early Alberta Mammals

The groups of mammals most familiar to us today first appeared during the Cretaceous Period, possibly evolving alongside the rapidly diversifying flowering plants. Marsupials, which today include kangaroos, koalas, and opossums, give birth to live, undeveloped young that must be fed and sheltered—usually in a pouch. Placental mammals, which include most modern mammals, are able to nourish and protect embryos within the bodies of mothers, allowing embryos to grow to an advanced stage before birth.

Known almost exclusively from fossil teeth, different groups of early mammals are identified by differences in teeth. Even today, mammal species can be identified by tooth shape and placement

Royal Tyrrell Museum

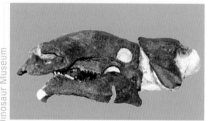

Fukui Prefectural Dinosaur Museum

This *Albertosaurus* (top) died young, 76 million years ago. Although scientists cannot say for sure what caused its early death, deformed bone growths from healing fractures on its right lower leg and in its foot may have played a part.

In 1998, the Royal Tyrrell Museum collected five ankylosaur skulls from Dinosaur Provincial Park—gaining, during the course of a few months, one of the world's largest collections of armoured-dinosaur specimens. Fossils of armoured-dinosaur skeletons (middle) are usually limited to bony plates that shielded the animals necks and backs.

Each species of horned dinosaur had its own distinguishing frill features and horn arrangements. *Centrosaurus's* frill (bottom) was scalloped around the edges, and small, curving horns decorated its top edge. Frill ornamentation may have helped individual centrosaurs recognize each other.

DINOSAUR GRAVEYARDS

Hiking through the heart of the park's palaeontological preserve in 1979, Dinosaur Provincial Park interpreters Ron Chamney and John Walper rounded a hill and stumbled into a *Centrosaurus* bonebed. The ground before them was so thick with dinosaur fossil fragments that they couldn't walk without stepping on bone.

Excavation of the site during the following decade yielded jumbled, broken evidence of more than 200 centrosaurs that had died in a flood 75 million years ago. As scientists traced the layer of rock in which the dinosaur graveyard is found, they discovered that it stretched across an area about the size of a football field. Other, similar bonebeds are clustered throughout the park, together representing thousands of centrosaurs.

Since the discovery of the *Centrosaurus* site, more than 200 other bonebeds have been identified throughout Alberta. Most contain a mix of fossils from different kinds of dinosaurs and other animals. Some, however, contain remains of a single dinosaur species, suggesting that the animals died together and may have lived together.

Royal Tyrrell Museum scientists David Eberth, Donald Brinkman, and then-palaeontology-student Michael Ryan surveyed many southern Alberta horned-dinosaur graveyards in the 1990s, looking for patterns in formation, distribution, and preservation. They discovered that species-specific bonebeds tend to be regionally clustered. One cluster along the South Saskatchewan River north of Medicine Hat contains 14 bonebeds scattered over a distance of four kilometres, which together probably contain the remains of thousands of individual centrosaurs. After tracking layers of coals and shales through the region, Eberth has suggested that the bonebeds in this cluster may have formed during a single disastrous event 75 million years ago—possibly a massive flood on the ancient coastal plain due to hurricanes sweeping in off the eastern sea.

By analyzing the position of clustered bonebeds in relationship to distinctive sediments which extend far beyond the limits of the clusters, Eberth, Brinkman, and Ryan believe horned dinosaurs ranged east–west throughout the year, not north–south as scientists had long thought. The animals may have migrated in huge herds to the drier western highlands for the rainy season, and returned to the wet coastal plains in the dry season.

Royal Tyrrell Museum

Fukui Prefectural Dinosaur Museum

A horned-dinosaur bonebed near Grande Prairie (top) contains remains of a group of horned dinosaurs called *Pachyrhinosaurus* (bottom). Six nearly complete skulls were found at this site—a rare occurrence in horned-dinosaur bonebeds.

Each species of horned dinosaur had its own distinguishing frill features and horn arrangements. Frill ornamentation may have helped individual centrosaurs recognize each other (bottom).

DUELLING BONEBED INTERPRETATIONS

The *Albertosaurus* bonebed in Dry Island Provincial Park, 80 kilometres up the Red Deer River from the Royal Tyrrell Museum, was discovered in 1910 by American palaeontologist Barnum Brown. He collected partial skeletons of nine albertosaurs from the site and shipped them to the American Museum of Natural History in New York. He then moved his explorations to what is now Dinosaur Provincial Park, and never returned to the *Albertosaurus* bonebed. The site was forgotten, and its location lost.

While examining meat-eating dinosaur remains in the collections of the New York museum in 1996, Tyrrell Museum Curator of Dinosaurs Philip Currie encountered the fossils Brown had collected. Using Brown's scant field notes and historical photographs taken of the American's field camp, Currie rediscovered the bonebed in 1997.

Since then, he has uncovered isolated remains of at least three more albertosaurs in the quarry, including one baby and one teenager. Currie believes the albertosaurs may have been living together as a family group when disaster struck 70 million years ago. "When animals—especially animals of such diverse age groups—die together, there is a good chance they may have lived together."

David Eberth, curator of sedimentary geology at the museum, disagrees. He believes an extreme climatic event occurring at the time of the animals' deaths brought the animals together. Climate, geology, and chance preserved what should have been a brief encounter for many million years.

Disagreement about interpretation of fossil evidence is not unusual among researchers; in fact, dissent is essential to scientific research. As Currie and Eberth collect more evidence, one or the other may prove his hypothesis, both may change their interpretations completely, or they may find entirely new questions that need to be asked to solve the mystery of the fossil record.

Philip Currie, Royal Tyrrell Museum

Of the approximately 200 known dinosaur bonebeds in Alberta, only the *Albertosaurus* bonebed contains almost exclusively fossils of meat-eating dinosaurs. Palaeontologist Philip Currie has discovered remains of at least 12 tyrannosaurs called *Albertosaurus* at the Dry Island Provincial Park site, including skull material from a teenaged and a baby *Albertosaurus*.

ANCIENT DINOSAUR YOUNG

Odd-looking fossil fragments found by teenager Wendy Sloboda in 1987 prompted the discovery of the first-known dinosaur-nesting site in Canada. Sloboda found the fossils along southern Alberta's Milk River Ridge. Thinking they might be fossil eggshell, she sent them to the University of Calgary to be identified. Researchers there confirmed her guess: the fossils were pieces of dinosaur eggshell.

Sloboda's discovery was well-timed. The Royal Tyrrell Museum of Palaeontology was planning an expedition to southern Alberta to look for dinosaur nesting sites, such as had been found across the border in Montana almost 10 years earlier. Within three days, the field crew was prospecting near the site of Sloboda's discovery, in a picnic spot called Devil's Coulee.

They spent several days searching the area. Although the crew found many fossil eggshell fragments, there was no trace of nests. Time was running short: there was one day left of prospecting scheduled before the crew returned to Drumheller.

Museum technician Kevin Aulenback paused in his search to rest on a hillside. As he sat down, he glanced down at the ground around his feet. Fragments of flat, textured pebbles were scattered about. Sharpening his focus, he picked one up. It was fossilized eggshell.

Prompted by his discovery, Aulenback quickly worked his way uphill to the place from which he thought the fragments were being washed down. In a layer of sandstone only metres above where he had been sitting, he found the source. Eggshells and tiny bones of unhatched dinosaurs were eroding from the hillside. Nearby, in the same rock layer, fossilized remains of other eggs, of entire nests, and even of young nestlings could be seen.

The first dinosaur-egg site known in Canada was found.

Seventy-five million years ago, duckbilled dinosaurs called *Hypacrosaurus* returned to the dry inland site, year after year, to build nests and lay eggs. Fossil evidence from this and other duckbill nesting sites in western North America suggests the dinosaurs were nurturing parents: they cared for their eggs and, after the young had hatched, may have brought food for nestlings. The herbivores would have quickly stripped the area of vegetation. Parents would have had to travel farther and farther for groceries, leaving growing babies vulnerable to predators such as *Troodon*, whose teeth have been found in a duckbill nest site near Drumheller.

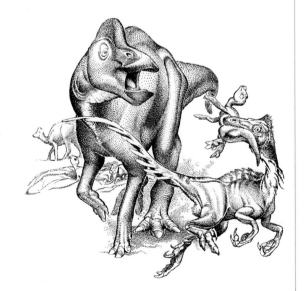

Seventy-five million years ago, duckbilled dinosaur *Hypacrosaurus* built nests, laid eggs, and raised young year after year at the same sites in what is now southern Alberta. *Troodon*, a small meating-eating dinosaur with a relatively large, well-developed brain, is thought to have eaten dinosaur eggs and hatchlings.

WHERE IN ALBERTA

More than 500 skeletons and countless bones of more than 35 Late Cretaceous dinosaur species have been found in Dinosaur Provincial Park since palaeontologists first started working there in 1898. The remains of other animals are equally well preserved: fish, salamander, crocodile, and turtle fossils are common in the park.

Volcanic ash contained in the park's rocks allowed scientists to identify the time when the sediments were laid down. Between 74 million and 76 million years ago, the region was part of an coastal plain cut by wide rivers that flowed from the growing western highlands. The area was green with vegetation.

High volumes of sand and silt carried into the region by the rivers permitted burial and fossilization of animal bones and other remains.

The rocks and fossils are exposed because glaciers and rivers later carved away rocks that had buried the dinosaur-fossil-bearing sediments.

In the ranchlands north of the Alberta–Montana border, dryness is a climatic theme which has endured a long time. The Milk River Ridge is cut here by Devil's Coulee, near Warner. The 75-million-year-old sandstones and shales that layer the badlands tell of a dry, forested plain cut by shallow rivers that seasonally flooded, dumped loads of mud and sand upon the land, and then dried up by summer's end.

This was the breeding ground for Alberta's *Hypacrosaurus*, whose nests, eggs, and babies erode from the coulee walls.

The Milk River Formation, exposed in the badlands of Writing-on-Stone Provincial Park, southeast of Lethbridge, records the first time that the region emerged from beneath the waves after the shallow seaway had flooded North America's interior. The rocks contain some of Alberta's oldest identifiable dinosaur bones and teeth. Duckbills, horned dinosaurs, meat-eating dinosaurs, and armoured ankylosaurs lived here 80 million years ago. The rocks also contain rare fossilized teeth of Cretaceous mammals.

Philip Currie, Royal Tyrrell Museum

Royal Tyrrell Museum

Dennis Braman, Royal Tyrrell Museum

While Dinosaur Provincial Park (top) is known around the world for dinosaur fossils that date from the Late Cretaceous Period, other places in the province, such as Devil's Coulee (middle) and Writing-on-Stone Provincial Park (bottom) are also rich with fossils from that time.

The Western Interior Seaway

Throughout its existence, from about 110 million years ago to about 70 million years ago, the inland sea that connected arctic waters to the Gulf of Mexico ebbed and flowed across Alberta. The broad plain between the rising western highlands and the eastern sea was flat. Slow-moving rivers, up to 50 to 125 metres wide and 10 metres deep, raised sand levees along their banks—rarely enough to contain storm surges that swept in from the coast more than 100 kilometres away. Mounds of sand, silt, and mud collected around the roots and trunks of trees during flooding in the same way that snow drifts around fences. These highpoints in the terrain sheltered dense thickets of shrubs and trees such as sycamores and sequoias—islands among the scrub that covered lower parts of the plain.

With land flat and low, the sea crept back and forth across the province throughout the Cretaceous.

About 74 million years ago, the water advanced over Alberta yet again, covering the 300-kilometre-wide coastal plain in less than one million years. It drowned forests and swamps, and turned what are now Medicine Hat and Lethbridge into one huge, shallow bay. The sea that formed is called the Bearpaw Sea, for Montana's Bearpaw Mountains, where scientists first recognized the sea's signature rocks.

Life in the Bearpaw was as diverse and hazardous as was life on land. Earlier in the period, a new kind of reptile predator, the mosasaur, took to the sea, joining ranks of seagoing reptiles such as plesiosaurs and marine turtles. Other marine carnivores included coiled- and straight-shelled ammonites, and sharks. Fish must have been abundant to support the predators.

About 72 million years ago, the Bearpaw began its long goodbye, retreating to the south and east, periodically pausing and retracing its path.

Predators in an Ancient Sea

The air scribe's whine fills the air. Palaeontologist Patrick Druckenmiller is in the lab at the Royal Tyrrell Museum, removing rock from the remains of a plesiosaur, a marine reptile that lived in the seas that covered northern Alberta more than 100 million years ago. It is one of nine specimens discovered between 1992 and 2000 by Syncrude Canada as the company mined northern Alberta's Athabasca oil sands.

The sands that now contain the heavy, tarry petroleum were deposited about 120 million years ago. At that time, northern Alberta was covered by a sea. Rivers flowed northwards across the province, carrying sand, silts, and clays. At the river mouths, the currents loosed their hold on the sediments, which settled in estuaries and along the shoreline.

Royal Tyrrell Museum

Patrick Druckenmiller (top) helped technicians at the Royal Tyrrell Museum prepare the Fort McMurray marine reptiles. He also determined what plesiosaur species were represented and how they might be related to one another.

Technicians preparing this short-necked plesiosaur fossil (bottom) found a number of marble-sized stones within the specimen's stomach area. These are believed to be gastroliths, or stomach stones—stones swallowed by the animal to help digest food or to provide ballast.

The rocks with the Fort McMurray marine reptiles lie immediately above the Athabasca oil sands. They formed 110 million years ago in shallow seawaters and mark the beginning of the northern sea's southward invasion of Alberta's interior. Specimens found in the mines include long-necked and short-necked plesiosaurs, and even a rare Cretaceous ichthyosaur known as *Platypterygius*.

"Good exposures of marine rocks dating from the Early Cretaceous are very rare," says Druckenmiller. "The Fort McMurray site contains both the best-preserved and the most-complete specimens known from that time on the continent."

AT THE BOTTOM OF THE WESTERN INTERIOR SEAWAY

C. J. Collom

Covering more than 1,000 square kilometres, the Clear Hills iron-ooid deposit is located in the Bad Heart Formation, seen here as the buff-coloured ledge rising from the water.

The Clear Hills stretch across fertile, rolling landscape north of Alberta's Peace River valley. Mixed woods of pine, spruce, poplar, and paper birch blanket the hills, hiding the most extensive deposit of iron ore in the province. Covering more than 1,000 square kilometres, the deposit contains more than one billion tonnes of iron ore.

The iron is in the form of peculiar, round grains of rock called "ooids" (rhymes with fluids). Each measures less than one millimetre across. By studying how these tiny, egg-shaped rocks form today, Mount Royal College palaeontologist Christopher Collom determined how the deposits of the Clear Hills region of Alberta formed 85 million years ago. Ooids develop in warm, shallow-water environments, such as those found among the islands of the Bahamas, where ocean water is supersaturated with calcium-carbonate minerals. Waves constantly roll grains of sand, pieces of shell, or other tiny particles back and forth on the seafloor. Similar to a rolling snowball, layers of clay and mud stick to the core and slowly build into a pellet. In the Bahamas, the underwater shelf and shoals are covered in limy muds; ooids that form there are made of calcium carbonate. In warm, shallow water near places where mineral broths seep out from deep within the Earth's crust, such as near some Indonesian islands, ooids are instead made of iron-rich clays.

The region of Alberta that is now the green, forested Clear Hills was covered by the Western Interior Seaway 85 million years ago. Collom believes it was also riddled with underwater mineral seeps. Cracks and fissures in the seafloor, reaching from deep within the Earth's crust, brought up brines of sulphur and iron minerals to the surface. The minerals settled from the solution as clays on the seafloor, providing the raw materials for the Clear Hills ooidal ironstones.

Syncrude

Syncrude

The Athabasca oil sands (right) cover an area in northeastern Alberta nearly twice the size of Lake Ontario. Scientists believe the petroleum either migrated upwards from Devonian Period-aged rocks immediately below or formed in other Cretaceous-rocks and was squeezed into the Fort McMurray sandstones (left), where it has lain trapped for millions of years.

Further research by Collom and by Royal Tyrrell Museum palaeontologist Paul Johnston indicates that many of the underwater mineral vents nourished complex communities of invertebrate animals that could harvest the chemicals and use them as food. Fossil bivalves called inoceramids present in vast numbers near mineral seeps in the region suggest that these animals used chemosynthetic bacteria in their gills to process hydrogen sulphide from seawater as a primary source of food.

Water circulation in Alberta's seaway is believed to have been limited, creating oxygen-poor seafloors. Inoceramids rested like rafts on the black, organic-rich, seafloor muds. Ranging from one centimetre in size to nearly two metres across, these giant clams often served as life rafts above the suffocating sediments for colonies of oysters and other small animals.

In 1998, Johnston and Collom rewrote the book on inoceramids with a single, controversial research paper. Once thought to be close relatives to oysters, the inoceramids are now considered to have belonged to a very different group of unique bivalves, the Cryptodonta. Johnston and Collom traced the existence of this group of clams deep into the early Palaeozoic Era, to the oldest known bivalved molluscs.

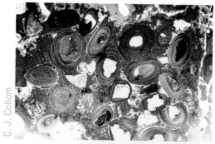

C. J. Collom

Paul A. Johnston, Royal Tyrrell Museum

Ooids (top) are tiny pellets that form in warm, shallow water when waves roll grains of sand or shell back and forth over fine mineral muds. Each ooid measures less than one millimetre across.

At almost two metres across, this inoceramid specimen (bottom) once lived at the bottom of the Western Interior Seaway, in northern Canada. The fossil was donated to the Royal Tyrrell Museum by Len Hills, of the University of Canada.

A New Look at Life in the Bearpaw

Royal Tyrrell Museum

Squirting a stream of water from its coiled shell, the ammonite propelled itself through the murky, half-lit depths of the sea. One more spurt and . . . from the surface sped a large shadow. Long narrow jaws opened and snapped around the spiral shell.

The mosasaur, ammonite clasped in its jaws, turned to the surface, sculling upward with its long, sinuous tail.

This is the traditional view of the relationship between mosasaurs and ammonites. The air-breathing marine lizard and the coil-shelled sea creature shared the world's oceans during the Late Cretaceous. Both were predators—ammonites hunted fish, as their modern squid and cuttlefish relatives do; and mosasaurs preyed upon ammonites. Fossilized ammonite shells are often found preserved with puncture patterns that match how teeth are positioned in fossil mosasaur jaws.

For years, scientists believed that punctures in fossil ammonite shells were evidence of eating habits of mosasaurs, a fearsome marine reptile of the Cretaceous Period (top). New evidence suggests some of the holes may have been caused by small, snail-like animals with cone-shaped shells, called limpets. A fossilized limpet is attached to this ammonite specimen (bottom).

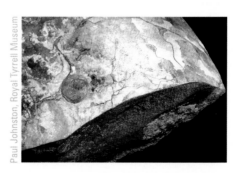

Paul Johnston, Royal Tyrrell Museum

However, Paul Johnston, of the Royal Tyrrell Museum, and Tomoki Kase, of Japan's National Science Museum, think that calling all circular holes in ammonite shells mosasaur bite marks is oversimplifying life in the Bearpaw. They think many of the holes were made by an animal much smaller and less threatening than fearsome mosasaurs. They think that other animal was the limpet.

Limpets are small, snail-like animals with wide, cone-shaped shells that are found today attached to rocks in intertidal areas. Limpets scrape away at the host surfaces they adopt, slowly creating round depressions called home scars. A limpet will move short distances away from its home scar to graze on algae, but will always return to it—or home in on it. Telltale scratches mark the animal's feeding paths. An ammonite specimen from South Dakota, now in the collection of Japan's National Science Museum, is infested with fossil limpets. Many circular depressions also decorate the specimen: some resemble home scars produced by modern limpets. Limpet scratches can also be seen.

According to Johnston and Kase, limpets colonized the shells along the waterline or along under-water lines marking temperature or water-chemistry layers. As the amount and distribution of water within a shell changed, so too would the shell's position within the surface of the ocean. The limpets migrated, settling in and creating new lines of scars.

The grazing-limpet theory explains how, on some specimens, double rows of punctures exist only on one side of the ammonite. If mosasaurs were solely responsible for creating the lines of holes by biting ammonites, the predators' upper and lower jaws would have made matching puncture rows on both sides of the shell.

As research continues, scientists may find that snacking mosasaurs are responsible for some punctured ammonite shells, and limpet hitchhikers for others.

In 1997, Paul Johnston and Tomoki Kase hauled a hydraulic replica of a mosasaur's jaws (top) to the Philippines and tested the effects of mosasaur bites on the ammonite's closest modern relative, the nautilus. They found that many of the shells cracked when the machine bit into them (right).

WHERE IN ALBERTA

The ancient marine reptiles unearthed by Syncrude Canada, in its Fort McMurray oil-sands mine, come from the Mannville Group of rocks, one of Alberta's oldest surviving records of the Cretaceous. These marine rocks mark a time when the province's rivers flowed north. In addition to skeletons of plesiosaurs and ichthyosaurs, the Mannville Group contains some of Alberta's richest fossil resources: the Athabasca heavy-oil sands of northeastern Alberta are the world's largest single oil-sand deposit.

A layer of rock in the cliffs near Drumheller marks the Bearpaw Sea's final farewell to Alberta, 70 million years ago. Known as the Drumheller Marine Tongue, the layer of golden sandstone is exposed at the top of the valley walls at Horsethief Canyon. The rock contains abundant oyster fossils.

In 1996, seven-year-old Lethbridge resident Andrew Morgan, sister Justine, and father Doug discovered fossil bones eroding from the bank of the Oldman River. They took some of the loose bones to the Tyrrell Museum, where staff identified the animal as a mosasaur. Morgan helped palaeontologists collect the skeleton's skull, backbone, ribs, and hips from the riverbank.

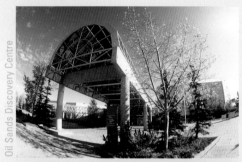

Oil Sands Discovery Centre

Dennis Braman, Royal Tyrrell Museum

Don Brinkman, Royal Tyrrell Museum

Fort McMurray's Oil Sands Discovery Centre (top) provides displays information about the region's oil-sands mines. Marine reptile fossils from the mines are displayed at the Royal Tyrrell Museum.

The Drumheller Marine Tongue (middle) at Horsethief Canyon contains abundant oyster fossils, while fossils of mosasaurs can be found in the Bearpaw Formation along the Oldman River (bottom).

MESOZOIC FINALE

As the Bearpaw Sea drained from Alberta, networks of marshes, forests, rivers, and deltas formed in its wake. Thickets of trees grew among lower-growing forests of ferns and horsetails on the plains. Trees, some reaching 30 metres towards the sky, were islands for other plants, which lived around their roots and in their branches. On higher ground, trees sheltered understoreys of ferns and shrubs. In lower areas with poor soil, swamp cypresses and flowering shrubs relied on floods to supply nutrients.

Reptiles such as champsosaurs, crocodiles, and turtles lived on the area's riverbanks. Birds competed with flying reptiles. Although new species had evolved in the few million years since dinosaurs had last lived in the region, the basic groups of duckbill and horned plant-eaters, and large and small meateaters were the same. As time passed, players in the dinosaur drama came and went: horned *Arrhinoceratops* gave way to *Triceratops*; fearsome *Daspletosaurus* was replaced by *Tyrannosaurusrex* .

As the Mesozoic Era entered its final few million years, the sea began to drain from North America. Climates cooled, and became drier and more extreme. Conditions favouring the formation of coal shifted south from the Northwest Territories and, as the Cretaceous wound to a close, as far as southern Alberta. Even then, however, there is no evidence of prolonged periods of frost in polar regions.

Ongoing research by Tyrrell Museum scientists into the preserved remains of the end of the Cretaceous Period indicates a steady lessening of diversity. Fossil pollen and spores, as well as remains of animals, suggest species were experiencing an increasing number of extinctions throughout the last few million years of the Mesozoic Era. While dinosaurs were getting bigger, as evidenced by *Triceratops* and its nemesis *T.rex*—two dinosaurs that reigned in the last years of the Age of Dinosaurs—they were also becoming less diverse. The Scollard Formation, the last group of Alberta rocks to form during the Cretaceous Period, contains about half as many kinds of meat-eating dinosaurs and about one-third as many kinds of plant-eaters as the Dinosaur Park Formation, which is 11 million years older.

Royal Tyrrell Museum

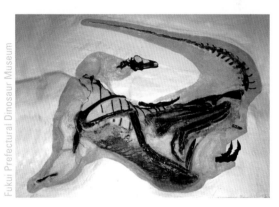

Fukui Prefectural Dinosaur Museum

Triceratops (top) is the largest known horned dinosaur, as well as one of the dinosaurs found in sediments closest to the end of the Age of Dinosaurs. It measured up to eight metres in length. Its metre-long horns probably helped protect it from predators such as *T. rex*.

Alberta's bird-mimic dinosaurs appear to have changed little throughout the Cretaceous Period. Five million years younger than the specimen discovered in Dinosaur Provincial Park in 1995, this *Struthiomimus* (bottom), found near Drumheller in 1997, is a smaller animal with a relatively smaller head.

Royal Tyrrell Museum

This *T. rex* specimen is named Black Beauty. The fossil's silvery-black colour is caused by the mineral manganese, which filled the bones' tiny pores and cracks during fossilization.

The final notes of the Age of Dinosaurs were performed dramatically by an asteroid. Sixty-five million years ago, an extraterrestrial bullet that was about 11 kilometres across struck the coast of the Gulf of Mexico at a speed of more than 200,000 kilometres per hour. It excavated a crater 80 kilometres across and more than 20 kilometres deep, and, in a matter of seconds, changed life on the planet. The impact vaporized the asteroid, seawater, and seafloor. Fire and molten pieces of the planet were flung into the atmosphere to rain down across North America, setting forests and plains alight. High-atmosphere winds carried smoke, ash, dust, and other airborne material to other parts of the globe, scattering the evidence around the world. On almost every continent, telltale signs of the explosion, its origins, and its aftermath can be found in a thin layer of clay that marks the boundary between the Cretaceous and Tertiary periods.

Many scientists believe that dust and ash from the impact triggered global acid rain, freezing temperatures, and long days and nights of darkness—adding to already-changing climates and environments of Late Cretaceous Earth.

For some animals, the impact and subsequent fast-forwarding of climate change was too much. Scientists currently estimate that 30 to 40 percent of all species that lived at the end of the Cretaceous Period disappeared during the impact event and its aftermath. Fossils of dinosaurs, pterosaurs, plesiosaurs, and mosasaurs are found below the boundary layer, but not above. Ammonites and some clams also became extinct.

Birds, fish, mammals, small amphibians such as frogs and salamanders, and small reptiles such as turtles, crocodiles, lizards, and snakes survived. When dawn did not arrive in the days, weeks, or months after the impact, small size, varied, adaptable diets, and protective habitats helped these animals to wait out the long, nuclear winter and repopulate the globe.

TRACING THE LAST DAYS OF THE CRETACEOUS PERIOD

Dennis Braman, Royal Tyrrell Museum

Dennis Braman, Royal Tyrrell Museum

A hillside north of Drumheller (top) documents the end of the Age of Dinosaurs and the start of the Age of Mammals 65 million years ago. Clay marking the time boundary (bottom) contains two layers. The lower, lighter-coloured layer may represent the sudden raining down of material forced into the atmosphere by the impact of an asteroid off the coast of Mexico. The darker layer above may be ash from the aftermath of the collision.

A hillside north of the Royal Tyrrell Museum exposes the last moments of the Mesozoic Era. Scientists have collected samples from the site, Canada's best exposure of rocks marking the boundary between the Cretaceous and Tertiary periods, since the early 1980s.

Dennis Braman, palynologist at the Tyrrell Museum, has spent decades analyzing the Cretaceous–Tertiary boundary. Working with evidence so minute that high-powered microscopes must be used, he and other scientists around the world examine the details of what happened just before, during, and immediately after the extinction event. They have taken the boundary apart, molecule by molecule, pollen grain by pollen grain. Contained within the boundary layer, they have found high concentrations of iridium, a platinum-like mineral rare on the surface of the Earth but common in meteorites. Altered quartz crystals, microscopic diamonds, and droplets of glass are other signs within the boundary layer that point towards impact by an asteroid.

Braman has also found evidence that impact by massive meteorite may not have been the sole cause of extinctions at the end of the Cretaceous Period. Immediately below the thin boundary clay layer is a zone of rock that ranges from one-half to two centimetres in thickness. Braman calls this the "impoverished zone." It should contain ample evidence of life in the form of fossil pollen and spores. Instead, it contains almost no fossils at all.

"Something big happened suddenly," Braman says. "Evidence supports an asteroid impact, but it also says something else was already going on."

Other research by Braman supports that conclusion. "The fossil record, here in the Red Deer River valley and elsewhere, clearly indicates a decline in the number of some plant groups during this time. Other groups of plants appear to have become more abundant and diverse—they conceivably were better adapted to whatever changes were going on."

TRACING THE LAST DAYS OF THE CRETACEOUS PERIOD

The last days of the dinosaurs can be traced in the walls of the Red Deer River valley (top)—one of the few places in the world where the last few million years of the Cretaceous Period are exposed (bottom).

And if plants were affected, so too were animals. The fossil record indicates that the diversity of dinosaurs was dropping at the same time. The question of what was causing the gradual extinctions has prompted scientists at the museum to investigate the final years of the Cretaceous. The research focusses on the walls of the valley in which the Royal Tyrrell Museum is located.

"The Red Deer River valley," says David Eberth, curator of sedimentary geology at the Royal Tyrrell Museum, "is one of the few places in the world where dinosaur-containing rocks representing the last few million years of the Cretaceous are clearly exposed. You can follow the river upstream and trace the passage of time right up to and past the boundary with the Tertiary."

By studying the sequence of rock layers in the valley walls, and the fossils contained within, the scientists have begun sorting out what was going on—climatically and environmentally—in the lead-up to the end of the dinosaurs.

Preliminary evidence collected from an *Albertosaurus* bonebed in Dry Island Provincial Park, 80 kilometres upstream of Drumheller, has convinced Eberth that weather patterns were changing dramatically 68 million years ago. The storms that he believes hammered the area may have contributed to the deaths of the bonebed's 12 or more tyrannosaurs, as well as shifting the balance between survival and extinction for many species.

It will be many years before the systematic survey of valley-wall sediments is completed. However, Eberth thinks answers to many questions scientists have about the last, most intriguing interval of the Age of Dinosaurs are hidden within Red Deer River rocks.

PRESERVED IN GLASS

In the Royal Tyrrell Museum's gallery, perfectly formed cones, needles, and twigs of conifer trees rest protected behind glass. They look as if they recently had been gathered from a redwood—one of the towering trees that grow along northern California's coast. If you peer closely, you can make out textures and patterns on the specimens. If you saw them through a microscope, you could trace the plants' cell walls.

These specimens came from a kind of redwood that lived in the Drumheller region 70 million years ago. The specimens are fossils. They are made of glass.

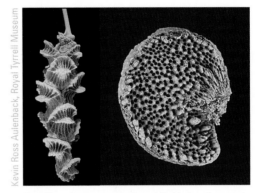

Fossil angiosperm seedhead (left), fossil *porosia* seed (right)

Not far from the Royal Tyrrell Museum in Midland Provincial Park is the quarry where the fossils were found. Seventy million years ago, it was a small pond. Cones, twigs, and needles fell into it from surrounding redwoods. Eventually, the pond filled with sediments. Other sediments collected above. Over time, they were preserved as rock within geology's landscape, keeping the plant material safe.

When an organism fossilizes, its organic molecules are replaced by nonorganic minerals. Microscopic holes and cracks within the remains may be saturated with groundwater minerals. In southern Alberta, the permineralizing compound is usually calcium carbonate. In the case of Drumheller's glass plants, silica, the substance that makes up quartz and glass, crystallized within the remains and preserved the cells' details.

Glass fossils of plants that lived in the Drumheller region 70 million years ago helped Tyrrell Museum technician Kevin Aulenback decide which plants to include in the Museum's Cretaceous Garden.

WHERE IN ALBERTA

The rich deltas, riverbanks, and backwater swamps that followed the retreat of the Bearpaw Sea from Alberta are preserved in the hills and cliffs of the Drumheller Valley.

Near the village of East Coulee, the transition from seafloor to river bottom can be seen at Hoodoos Recreation Area. The red-brown bases of the hoodoos represent seafloor muds of the Bearpaw Sea. They are topped by columns of white sandstone, laid down as beach sands 72 million years ago. The hard cap rock is sandstone deposited in the delta and tidal flats.

Lush swamps separated from the sea by the river levees have since become the beds of coal that stripe the valley walls. In the early part of the century, almost 150 coal mines operated in the 50-kilometre valley, from East Coulee to Nacmine, supporting a population of more than 50,000 people. Production peaked during World War II, when 40 trains, each pulling more than 80 railcars piled high with coal, lumbered from the valley onto prairie level every day in attempts to feed wartime industry.

Although oil and gas were being collected from Alberta as early as 1911, the discovery of the immense oil fields beneath Leduc and Turner Valley in the late 1940s led to the decline of Drumheller's mines. The area's last operating coal mine, the Atlas Coal Mine, closed in 1979. It is now a historic site.

In the early 1980s, a hillside near Huxley, Alberta, surrendered a partial skeleton of *Tyrannosaurus rex*. Alberta's other *T. rex* fossil comes from farther afield. In 1982, Black Beauty was collected from along the Crowsnest River east of Lundbreck.

Royal Tyrrell Museum

Dennis Braman, Royal Tyrrell Museum

NA-2389-67, Glenbow Archives

Royal Tyrrell Museum

Strange, mushroom-shaped formations called hoodoos (top) expose rocks made from sediment laid down during different environments in the Drumheller Valley's past.

In the early part of the twentieth century, almost 150 mines pulled coal from the hills around Drumheller (middle-top). Today, the last surviving mine, the Atlas Coal Mine, is a historic site (middle-bottom).

High above the Crowsnest River (bottom), rocks that date from the end of the Cretaceous Period yielded a fossil skeleton of *Tyrannosaurus rex*.

MAKING WAY FOR MAMMALS: THE CENOZOIC

Now there were no dinosaurs. Wiped out in the Cretaceous extinction, the animals that had dominated the world's landscapes for almost 150 million years were not part of Tertiary Period landscapes. Also vanished were marine and flying reptiles, ammonites, some clams, and two of the three families of marsupial mammals. Once again, in ecosystems around the world, vacancy signs went up.

Royal Tyrrell Museum

Tim Schowalter

Many animals that live in Alberta today, including prairie falcons (top) and Richardson's ground squirrels (bottom), are descended from creatures that first appeared during the Age of Dinosaurs.

FAMILIAR GROUND

The air never stops moving atop the Porcupine Hills. Forest envelopes you in a rush of sound and shifting light. The spruce trees are both masts and sails on this rocky promontory above the prairie: the long trunks sway in the wind, the sturdy green branches flap and twist.

You can see a long way from atop these hills. To the west, the Rocky Mountains march northward, a 1,500-metre barrier rising at the edge of the plains, over which massive sky currents swirl. To the south, rolling hills and tree-lined coulees recede, carved by ice, water, and weather. To the east, grass-covered plains rise and fall, swells on a petrified ocean.

The Porcupine Hills are a recent feature on Alberta's landscape, but elements of the scene remain unchanged from the time when dinosaurs disappeared. Even then, highlands—low compared to today's Rockies—marked the province's western horizon and funnelled wind over their ramparts. Conifers and broadleaved trees dotted the ancient plain that was Alberta near the end of the Cretaceous Period. Mammals were small and shrewlike—more similar to insectivores than to ground squirrels. And, today, the last of the theropod dinosaurs, the birds, travel the sky.

The 65 million years that separate us from the end of the Cretaceous encompass the Cenozoic Era—the time of Recent Life. Since the Cenozoic Era's dawn in the aftermath of an asteroid strike, Alberta has assumed the appearance and residents we know today.

The Cenozoic Era contains two geological periods. The 63-million-year stretch of time at the start of the Cenozoic is called the Tertiary Period, although some scientists prefer to divide it into the Palaeogene and Neogene periods. The final two million years of the Cenozoic make up the Quaternary Period. These periods are divided into seven smaller units of time, called epochs.

Like other ancient Alberta hilltops that escaped glaciation during the Ice Age, the highest points of the Porcupine Hills witnessed many changes during the Cenozoic Era.

Dennis Braman, Royal Tyrrell Museum

TERTIARY TRANSITIONS
Survivors and Successors

Alberta quickly recovered from events that transformed the world at the end of the Mesozoic Era. Temperatures warmed; heavy annual rainfall returned. Where the Western Interior Seaway had flooded the province, forests of conifers and broadleaved trees colonized the land, separated by rivers, lakes, marshes, and open, brush-filled plain. During the first few million years of the Cenozoic Era, Alberta once again assumed the environment that it had enjoyed during the latter half of the Age of Dinosaurs.

But now there were no dinosaurs. Wiped out in the end-Cretaceous extinction, the animals that had dominated the world's landscapes for almost 150 million years were not part of Tertiary Period landscapes. Also vanished were marine and flying reptiles, ammonites, some clams, and two of the three families of marsupial mammals. Once again, in ecosystems around the world, vacancy signs went up.

Survivors included birds, fish, turtles, lizards, snakes, and frogs. Mammals benefited most from the disappearance of dinosaurs. Being both generalists and opportunists, they adapted easily to new environments and new diets. Some mammal groups had had a head start on diversification by moving into vacant niches in the final years of the Cretaceous Period. When the asteroid reset the planet's evolutionary clock 65 million years ago, they were already running full stride in the race for dominance among survivors.

Fukui Prefectural Dinosaur Museum

Xian-chun Wu

d'histoire naturelle, Paris. Ref. Gardner, J. D. 2000

Scientifiques de la Musée nationale

Dinosaur Museum

Fukui Prefectural

At the end of the Cretaceous Period, turtles (top) were one group of animals to win the race for survival. If an asteroid that measures 11 kilometres across strikes the planet, it does not matter how big you are or how fast you can run. In those rare instances, it actually pays to be small, to live in or near water, and to eat other small animals—as turtles do.

Alberta's oldest Cenozoic Era vertebrate fossil is a crocodile (middle-top). It was found about 50 kilometres from Drumheller, just above the line of clay that marks the boundary between the Cretaceous and Tertiary periods. Crocodiles lived mostly in or near wetlands during the Age of Dinosaurs. During the Cenozoic, some groups of crocodiles moved from ponds and rivers to the sea.

The oldest fossils of albanerpetontids (middle-bottom), a group of animals related to frogs and salamanders, date from the Jurassic Period. After diversifying throughout the Cretaceous Period, a few of the amphibians survived the extinction that killed the dinosaurs only to disappear about 50 million years ago.

Although they resemble crocodiles, champsosaurs (bottom) are a Cretaceous-aged member of a much older group of reptiles. Fossils are common in Alberta rocks that date from the Late Cretaceous. Champsosaurs, survivors of mass extinctions at the ends of the Triassic and Cretaceous periods, disappeared about 50 million years ago.

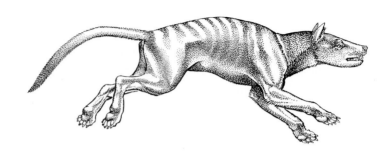

The earliest members of many modern mammal groups became common shortly after the dinosaur extinction. Primitive primates, ancestors of hoofed mammals, and insectivores related to modern hedgehogs and moles may have been around when dinosaurs breathed their last: these mammals are known from the Palaeocene Epoch, the span of time covering the first 10 million years of the Tertiary Period. Just 10 million years later, rodents, whales and sea cows, bats, true carnivorous mammals, and ancestors of camels, deer, and bison had joined them. Within 25 million years of the start of the Cenozoic, the number of mammal families had grown from the three or four groups that had survived the Cretaceous extinction to more than 75 families. The greatest diversification occurred within the group of mammals that bear young within their bodies for long periods before giving birth—the placentals.

Predators of early Cenozoic mammals included lizards, crocodiles, and large, flightless, carnivorous birds. However, mammals such as creodonts (top, left) and miacid carnivores (top, right) themselves soon filled that niche as well. They eventually were replaced by the ancestors of modern cats, dogs, weasels, bears, and seals.

Early ancestors of modern hoofed mammals had generalized teeth. Condylarths (bottom) were primarily browsers, filling niches vacated by duckbilled and horned dinosaurs at the end of the Cretaceous Period. Most ranged from hedgehog-size to dog-size.

TURTLE DIASPORA

Turtles were little affected by the mass extinction at the end of the Cretaceous, but they began a slow turnover during the Tertiary Period. Royal Tyrrell Museum palaeontologist Don Brinkman specializes in the study of Alberta's ancient turtles and their evolutionary relationships with fossil turtles found in other parts of the world. His research shows that, during the Tertiary, turtles trod the path taken by many dinosaurs that had lived in Alberta during the Late Cretaceous: Asian turtles invaded North America by way of the landbridge that periodically linked Alaska to Siberia from the Cretaceous Period onwards.

Groups of turtles that had first evolved in Alberta and had survived the extinction at the end of the Cretaceous disappeared around the end of the Palaeocene Epoch, about 50 million years ago. Invaders claimed turtle niches throughout the continent. Many modern turtles native to North America today are descendants of the Asian conquerors.

Don Brinkman, Royal Tyrrell Museum

Royal Tyrrell Museum

Plesiobaenea (top) was one of the Asian turtles that invaded North America in the early Cenozoic Era, displacing North American turtles such as *Basilemys* (bottom). These two *Basilemys*, found fossilized together in Dinosaur Provincial Park, were killed during a volcanic eruption.

OVERLOOKED SURVIVOR

Therapsids, mammal-like reptiles, were far more tenacious than scientists had thought. A jaw and isolated teeth of one such reptile were discovered by University of Alberta field crews near Cochrane in 1988.

Therapsids are primitive animals on the evolutionary line to true mammaldom. They dominated the animal kingdom on land during the Triassic Period. For years, scientists had believed the group died out during the Jurassic. The youngest such fossil dates to about 180 million years ago.

The Cochrane fossil dates from 60 million years ago, proving that this animal group survived not one, but two major extinction events.

2002 Museums and Collections Services, University of Alberta.

The therapsid jawbone and teeth collected near Cochrane in 1988 proves these mammal-like reptiles survived two major extinction events.

WINDOW ON A POST-DINOSAUR WORLD

Palaeontologists and geologists read rocks as if they were stories in a book: each kind of rock forms in a specific environment; each represents a chapter in the geological history of an area.

East of the city of Red Deer, a road cut along a highway exposes the changing history of early Cenozoic Alberta. Thin layers of exposed rock tell the story of the slow transformation of an early Cenozoic floodplain into a swamp flooded by water and sediment from a nearby river about 60 to 55 million years ago. Preserved within leaves of rock at Joffre Bridge are the fossilized characters that lived in the ancient environment: plants, freshwater molluscs, primitive mammals that range in size from a dozen or so centimetres long to as large as sheep, insects, wetland plants, and spawning fishes. Together, the rock layers and fossils show a succession of wetland environments, from floodplain to river to pond to swamp.

During the last 25 years, palaeontologists have come to the site to read the rocks and collect fossils, trying to understand the ancient communities that lived there and how they and the environment changed through time. Palaeobotanist Ruth Stockey and student Georgia Hoffman of the University of Alberta drew together existing research on the geological story of the area in 1999 by cataloguing rocks and plant fossils found at the site.

The change in fossils at the site parallels the change in sediments that form the rocks. Floodplain mudstone, clays, and coals contain fossils of roots, molluscs, small mammals, and reptiles. In the sandstone, mudstone, and siltstone that formed during the site's river phase, leaves, fruits, and seeds of river-valley trees and shrubs are preserved. When the river channel moved, fine sediments collected within an oxbow pond, forming siltstone, sandstone, mudstone, and clays. Fossils of river-valley tree seedlings and germinating seeds are preserved in the fine-grained sediments, as are fish scales and impressions of dragonfly wings.

The chapter describing the area's time as a swamp is told by coaly mudstone, freshwater mollusc shells, fish bones, and the twigs and cones of bald cypresses and sequoia-like trees. Gradually coarsening mudstones, siltstones, and clays represent repeated flooding of the swamp by the nearby river. An increase in the num-

Dennis Braman, Royal Tyrrell Museum

Joffre Bridge allows scientists to trace how one particular site changed through time during the Tertiary.

ber and kinds of fossils may reflect both plants that grew in the area and parts of plants that were carried in by floodwaters. Rootless aquatic plants such as duckweed, water ferns, and liverworts grew in the new environment, as did swamp-loving conifers. Other fossils include leaves of shrubs that resemble dogwood, sycamore-like trees, horsetails, ferns, and gingers, as well as shells, fish scales, and impressions of dragonfly, caddisfly, and beetle wings. This last phase in the rock-recorded history of the site contains the skull and part of the body of a large early Cenozoic mammal, called a pantodont, along with one of the most remarkable and rare finds of fossils of river fishes in Canada.

Almost 2,000 specimens of 57- to 55-million-year-old fishes are preserved in a layer of silty mudstone that measures only centimetres thick. Most of the fossils belong to the trout-perch family, while dozens are primitive smelts. Two specimens are relatives of bony-tongue fishes that now live only in Africa, South America, and Southeast Asia. The majority of the fishes are preserved in groups, facing in one or two directions, suggesting that they all died together. Some show fin structures known from breeding fishes; perhaps they were trying to spawn in shallow water when they died.

University of Alberta fish-palaeontologist Mark Wilson mapped the fossils. He believes the animals may have been trapped after a nearby river flooded its banks and spilled into the swamp during the spring spawning season. The bodies were buried in river-borne sediments soon after: most of the skeletons are intact and undisturbed by scavenging.

While remains of fishes are fairly common in ancient lake or marine deposits, they are rare in river sediments. Currents pull delicate fish skeletons apart, leaving only water-worn fragments of denser bones and teeth. This skews scientists' knowledge of fishes that lived in ancient rivers: much more is known about large, heavy-boned predators than about the smaller, more delicate species, such as the trout-perch and smelt found at Joffre Bridge.

Royal Tyrrell Museum

A pantodont was a heavy-boned, bear-sized mammal of the early Cenozoic Era. Two specimens have been discovered at Joffre Bridge.

Mark V. H. Wilson

Almost 2,000 fish specimens are preserved at Joffre Bridge, including fossils of trout-perch fishes (top) and primitive smelts (middle), as well as *Joffrichthys* (bottom), a relative of today's bony-tongue fishes.

Evolving Earth: The Cenozoic

By about 20 million years ago, the continents had assumed a modern configuration across the face of the planet. Final adjustments affected ocean circulation, which would in turn influence climate —contributing to the cooling of the Northern Hemisphere during the latter part of the Cenozoic Era.

The building of the Rocky Mountains changed Alberta dramatically during the Cenozoic Era. The slow accumulation of island chains against the prow of North America that had started in the mid-Jurassic neared its modern state. During the 30-million-year Cenozoic battle between the oceanic plate and North America, most of the floor of the western Pacific Ocean was recycled under the continent. Stacks of rock, each more than three kilometres thick, piled eastward. By about 35 million years ago, when the Eocene Epoch ended, their broken, jagged edges formed the tops of mountain ridges along the Alberta–British Columbia border.

The Rocky Mountains had formed.

Once in place, the great, stony spine of Alberta determined the province's geological and environmental fate. It drained the dregs of the Bearpaw Sea from the prairies. It increased the rate at which sediments were redistributed within the province; rivers and streams careened down from the peaks, cutting away at rock and terrain. On the plain, their journey slowed, and they marked their passage by depositing loads of sediment along their routes.

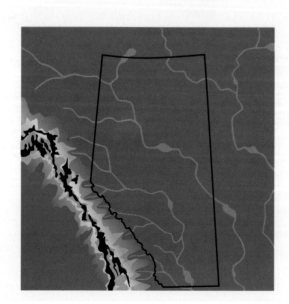

During the first part of the Cenozoic Era, rivers flowed north from the mountains across the province, but the great continental ice sheets of the Quaternary Period rerouted rivers southward.

The mountains also changed weather. Siphoning moisture from the winds that blew east from the Pacific Ocean, the Rockies created a 700-kilometre-wide rain shadow over the plains. Cool, drier weather prevailed. Wetlands dried, and the evergreen-and-broadleaved forests that carpeted Alberta retreated, giving way to drought- and cold-resistant grasses. By 35 million years ago, Alberta no longer resembled the hot, humid southeastern United States; climate and landscape were similar to modern African savannahs, with grassy plains dotted by small, open woodlands.

Such geological and climatic changes were not limited to North America. Around the world, once-isolated continents were getting reacquainted, with sky-high sutures marking their joining. India's collision with Asia began about 55 million years ago and continues today: the children of the union, the Himalayas, pierce the sky at four kilometres above sea level. When Africa bumped into Europe, mountain chains leapt up in North Africa, the Adriatic, and Italy. Elsewhere, Australia moved away from polar Antarctica, and Saudi Arabia drifted apart from Africa, tearing the crust and creating geological hot spots on their edges.

Rearrangement of continents affected ocean circulation. New mountain ranges influenced regional climates. Ice sheets formed in polar regions and at high altitudes. In the Northern Hemisphere, short, cool summers and long, mild winters became normal. Winter snowfall exceeded summer melt: glaciers and ice packs grew, creeping forth from mountain and polar strongholds. The first ice sheet covered North America and Europe almost two million years ago in the Pleistocene Epoch.

Advancing and retreating, ice blanketed parts of Alberta four times during the Quaternary Period. Continental glaciers up to two kilometres thick bulldozed the land, scraping away the surface of the province and transforming it into a barren sea of frozen water.

The rise of the Rocky Mountains in the west determined Alberta's climatic fate throughout the Cenozoic Era. With less rainfall, forests retreated to the north and grasslands became established.

As the floor of the Pacific Ocean was forced beneath North America, collisions fractured the continent's crust, causing volcanoes to appear in British Columbia's interior and further buckling western North America.

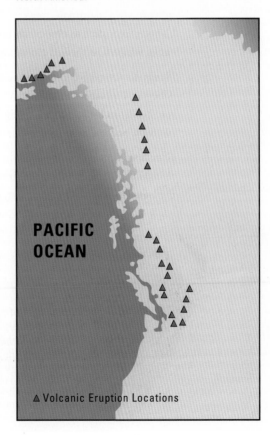

PACIFIC OCEAN

△ Volcanic Eruption Locations

Turning Point

Dennis Braman, Royal Tyrrell Museum

While mammals diversified, flowering plants dominated Alberta vegetation. Evolving hand in hand with mammals, flowers became more complex, and pollen, fruits, and seeds, bigger. Canopy trees such as elm, birch, oak, alder, and maple joined the ranks of sycamores and katsuras, which had thrived alongside dinosaurs.

Plants do not evolve in isolation of geography and climate. The Rocky Mountains influenced Alberta's vegetation 35 million years ago just as they do today. The cool, drier weather that resulted from the mountains' rain barrier caused the retreat of the great mixed-wood forests, swamps, and marshes of the Palaeocene, and the establishment of grass across the plains.

Grass's success is due to a combination of qualities, growth habits, and animals that evolved alongside. Unlike most plants, grass does not grow from the tips of its shoots; it grows from the base. Its roots are the largest, most extensive part of the plant, forming dense, tangled,

Dennis Braman, Royal Tyrrell Museum

It is difficult to imagine Alberta without grass, but until about 55 million years ago, grasses did not exist, and until about 35 million years ago, they grew only in small, limited areas. Now, grass covers one-third of the Earth's land surface, including one-third of Alberta (top). It is the most readily identified kind of vegetation in the province.

As the Cenozoic Era progressed, new forms of broadleaved trees (left) dotted the landscape and lined river valleys.

Royal Tyrrell Museum

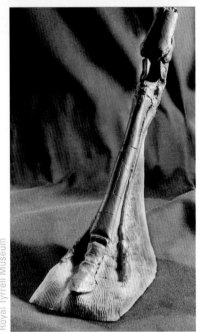

Royal Tyrrell Museum

underground mats. This investment in subsurface structure means that grass survives cold, drought, intense harvesting, and even fire. Healthy grasslands require periodic grazing—or burning or other intense disturbance if grazing is not possible—to eliminate shrubs and other plants that would crowd out the grass and change the ecosystem.

Dependence on disturbance suggests grass and grazing animals evolved together. Leaves and stems of grass contain silica, the same mineral that forms quartz, one of the hardest known rocks. Herbivores that make their living by grazing require special equipment. A mouthful of hard, high-crowned, grinding molars, strengthened by folds of enamel, allows grazing animals to chop and chew tough grass leaves and stems while preventing teeth from wearing down completely during an animal's life. New digestive equipment in the form of extra stomachs and friendly gut bacteria helps to extract nutrients. And, to outrun hungry predators on the open plains, grazers developed long legs and speed. Some later grazers, such as plains bison, were prairie-disturbance machines: their hooves were sharp, and they cut and tore plants as the animals moved across the plain. Fur traders' reports tell of the vast herds of bison leaving nothing but dust and hoofprints their wake. After such devastation, grass would quickly resprout from protected roots to renovate the prairie.

Condylarths, hedgehog- to rhino-sized herbivores of the Tertiary Period, were the genetic source for many modern lawn-mowing mammals. Odd-toed ungulates include horses, tapirs, and rhinos, as well as now-extinct brontotheres—elephantine grazers with slingshot-shaped nose horns. Other condylarth descendents are even-toed ungulates, which include pigs, hippos, cattle, deer, giraffes, camels, and antelope.

As grasslands and grazers evolved and spread, changes in other animal populations kept pace. By 35 million years ago, all major modern mammal groups were established and had crowded out less-successful kinds. Carnivores diversified into weasels, cats, dogs, and bears. Rabbits, apes, and monkeys were among the last groups to appear.

The grinding teeth of grazers (top) are covered in hard enamel that is folded into the structure of the teeth to provide support and resistance to wear.

The evolution of the horse—from forest-dwelling, terrier-sized *Mesohippus*, whose 30-million-year-old fossils have been found not far from Alberta's eastern border, in Saskatchewan's Cypress Hills, to the wild ponies that herd on Central Asia's steppes—reflects the ecological history of the Tertiary Period. *Hyracotherium* lived in forests and ate leaves from bushes and trees. As forests declined and grasslands spread, its descendants increased in size, reduced the number of toes on each foot to one (bottom), lengthened their legs, and developed teeth suitable for chewing grass.

WHERE IN ALBERTA

Rocks dating from the earliest years of the Cenozoic Era can be found striping the walls of the Red Deer River valley above older, Cretaceous-aged sediments. The Scollard Formation north of Drumheller contains fossils of early mammals, primitive trouts, crocodiles, and flowering plants. However, the younger Tertiary rocks in Alberta are, the rarer they are. Erosion during the late Tertiary stripped most of the rocks from the province. Only scattered remnants remain. These late Tertiary oases loom over the plains, capping the province's larger hills.

Southeastern Alberta's Cypress Hills are the highest point of land between the Rockies and Labrador. The crown of the hills consists of gravels and stones that broad, fast-flowing rivers carried from Montana's Sweetgrass Hills and Bearpaw Mountains. Laid down along riverbeds and banks, the gravels eventually hardened into a kind of mixed rock called conglomerate. Hidden under the yellow-coloured, natural cement are fossils of animals that lived in the area between 44 million and 35 million years ago, during the Eocene Epoch. Remains of birds, snakes, turtles, amphibians, and fish attest to the groups of animals that crossed over the Cretaceous–Tertiary boundary and thrived in the Cenozoic. Across the provincial boundary in Saskatchewan, the Cypress Hills are rich in mammal fossils, including rodents, flying lemurs, bats, rabbits, primitive horses, rhinos, camels, antelope, and giant brontotheres.

Other early Eocene sites cap Alberta's Hand Hills, the Swan Hills, the Saddle Hills, and the highest points of the Porcupine Hills.

The rising of the Rocky Mountains tortured Alberta's ancient rocky floor. About 50 million years ago, the crust deep beneath the province fractured from the stress created by the pushing, piling rock sheets that formed the Rockies. Magma seeped into the cracks and hardened before reaching the surface. The Milk River dykes, east of Writing-on-Stone Provincial Park, emerged as vertical black fortresses of rock after the surrounding soft bedrock was worn away.

Southern Alberta's highest hilltops contain remnants of rocks that formed during the late Tertiary. Beneath the cementlike conglomerates atop the Cypress Hills (top) are fossils of animals that lived between 44 million and 35 million years ago; and in the gravels and sands at the summit of the Hand Hills (middle) are remains of rodents, camels, and horses.

The Milk River dykes (bottom) chronicle how the formation of the Rocky Mountains fractured Alberta's basement rocks, allowing magma to seep into the cracks.

QUATERNARY COLD
The Big Ice

Ice ages—extended periods during which ice covers all or large parts of the Earth—are uncommon. Ice-gouged rocks and glacial debris within the geological record speak of global glaciation towards the end of the Precambrian, of ice sheets burying southern continents at the end of the Ordovician, Devonian, and Permian periods, and of giant glaciers repeatedly freezing the Northern Hemisphere during the last two million years. For most of Earth's history, however, the planet has been ice-free.

According to the fossil record in Canada's Arctic, the North Pole experienced only infrequent freezing temperatures during the 150-million-year Age of Dinosaurs. During one particularly warm period, about 80 million years ago, climate-dependent reptiles such as turtles and champsosaurs thrived in the north. Royal Tyrrell Museum palaeontologist Don Brinkman, one of the scientists who studied the fossils, says they indicate that the northern climate resembled that of South Carolina, with temperatures ranging from about 6° to 25° Celsius; turtles and crocodiles cannot withstand prolonged periods of severe cold. Brinkman and his colleagues believe increased levels of greenhouse gases in the atmosphere, due to a surge in volcanic activity around the world, may have caused the warm temperatures.

However, about 45 million years ago, the planet began cooling in earnest. Scientists believe a number of astronomical and geographical factors may have triggered processes resulting in the two million years of Northern Hemisphere deep freezes that we call the Ice Age.

Slight variations in the tilt of the Earth's polar axis influence seasons and climate on the planet's surface. The planet's tilt causes seasons by varying the intensity of sunlight shining on a region of the planet throughout a year. The Earth, however, wobbles; during the course of a 41,000-year cycle, the angle of tilt ranges between 22 degrees and 25 degrees, increasing or decreasing seasonal extremes.

Because the planet's orbit is slightly oval in shape, one pole is always closer to the Sun during its winter, while the other pole is closer during its summer, making for warm winters and cool summers in one hemisphere and more extreme seasons in the other hemisphere. A wobble in the orbit alternates the hemispheres' mild- and extreme-seasonal cycles about every 20,000 years. A larger variation in the orbit, called eccentricity, alternately lengthens and shortens the Earth's elliptical path around the Sun—strengthening or weakening seasonal extremes on the planet.

Kevin Ross Aulenback, Royal Tyrrell Museum

Although global cooling is documented in the fossil record around the end of the Age of Dinosaurs, temperatures rebounded shortly after. A fossilized palm-tree frond found in Cenozoic rocks near Genesee speaks of year-round, frost-free temperatures in the region 60 million years ago; palms die when exposed to prolonged freezing temperatures.

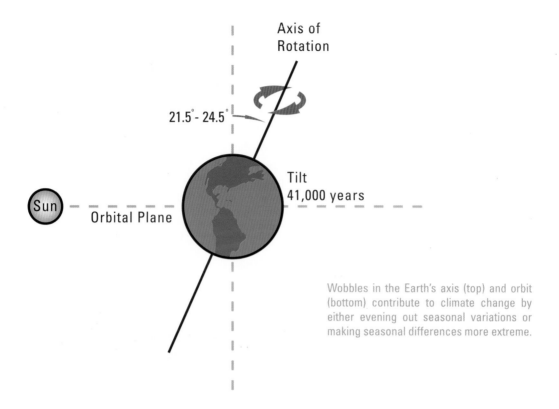

Wobbles in the Earth's axis (top) and orbit (bottom) contribute to climate change by either evening out seasonal variations or making seasonal differences more extreme.

The arrangement of continents on the planet's surface influences regional and global weather patterns. During the Tertiary Period, continents isolated the poles. Antarctica is barricaded by an uninterrupted stream of strong westerly winds that redirect warm ocean currents flowing from the equator. With the continent positioned over the South Pole, ice formed in the region. Ice absorbs less heat than land or open water, allowing antarctic glaciers to perpetuate themselves by reflecting sunlight back into space. Now, snow and ice cover most of Antarctica to depths of several kilometres, with only mountain peaks poking through.

The Arctic Ocean is isolated from equator-warmed ocean currents by a ring of continents and islands. Because most of Canada's fresh water drains northward via the McKenzie River and Hudson Bay, and many Russian and Siberian river systems end in the Arctic, water in the northern ocean is less salty than most ocean water, and it freezes more easily. Permanent floating ice covers the North Pole, reflecting warming solar energy back into space and further cooling the region. The ice pack grows during winter months.

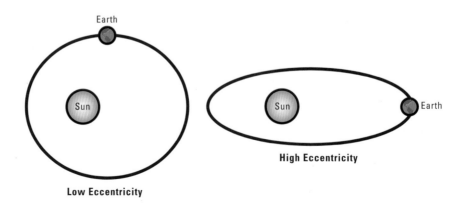

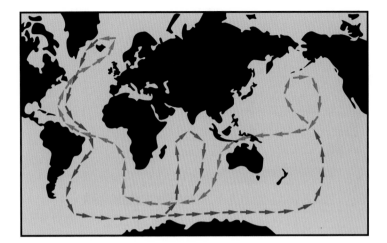

Ocean currents influence global temperature by controlling the flow of water warmed near the equator around the globe and by mixing salt water with fresh water, which freezes at higher temperatures.

→ **Warm ocean currents**
→ **Cool ocean currents**

When Central America closed the gap between North America and South America about 4.5 million years ago, it further complicated ocean currents by blocking the flow of water between the Pacific and Atlantic oceans. Equatorial water is redirected northward, bringing more precipitation to North America and Europe, which falls as snow in cold regions. More warming solar radiation bounces off the white surfaces into space. At times, annual snowfall outpaces annual snowmelt—the recipe for glacier formation. During glacial periods, the hemisphere becomes a self-fuelling, ice-forming machine.

Beginning about 2.5 million years ago, the chilling of the planet accelerated, heralding the start of the Quaternary Period, the most recent geological period in Earth's history. Punctuated by recurring ice, the Quaternary saw the final transformation of ancient geography and life into the landscapes, plants, and animals of modern Alberta.

Ice covered regions of the Northern Hemisphere more than 17 times; Alberta was buried four times. Ice sheets born near Hudson Bay expanded southward and westward to meet immense glaciers flowing eastward from Rocky Mountain valleys. Together, they re-landscaped the terrain, scraping away vegetation, grinding up rocks, and gouging bedrock. The debris—trees, boulders, gravel, and sand—was swept up into the ice and carried thousands of kilometres.

North America's greatest ice sheet, the Laurentide, covered more than 13 million

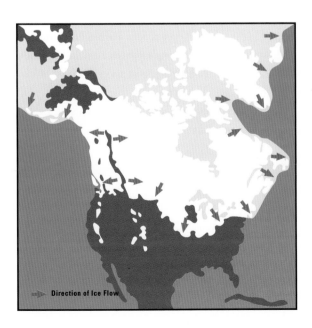

Direction of Ice Flow

The ice sheets that buried Alberta during the Quaternary Period originated from two places: from the north near Hudson Bay and from the west among the Rocky Mountain glaciers.

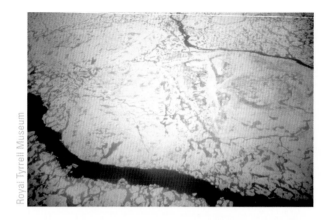

Royal Tyrrell Museum

White ice reflects more of the Sun's energy than dark, open water, crating colder temperatures and encouraging more ice to form. Because of this, a floating ice pack covers much of the Arctic Ocean throughout the year.

square kilometres. It reached as far as Montana, Minnesota, and New Jersey. The glaciers blocked valleys. Rivers and streams that had flowed to the east and north from the mountains during the late Tertiary Period cut southward-flowing channels along the edge of the ice.

As the ice melted, runoff and blocks of stagnant ice submerged 90 percent of Canada's prairies. Meltwater formed huge lakes, dammed by ice and glacial debris. The alga-rich clays that collected on the lake bottoms show themselves today as rich soils such as those found in the Edmonton area. Water inevitably worked ways out of the lakes: the outflow, following paths of least resistance, found existing river channels. Water choked with mud and ice floes as big as three-storey buildings charged through the spillways, carving wide, deep channels into the underlying rocks. They created and filled canyons up to three kilometres across and a half-kilometre deep. In the few days it took for Glacial Lake Drumheller to drain, several thousand cubic kilometres of rock and sediment were stripped from the plains. Glacial Lake Drumheller was only one of many such lakes found in Alberta at the end of the last ice age.

Just as scientists cannot say for certain what triggers the spread of continental ice sheets, neither can they say what causes the 10,000- to 25,000-year warm periods within ice ages. Albertans live in one such period today.

Off the Ice

The continents' latest glacial period lasted from about 22,000 to 12,000 years ago. Throughout most of it, Alberta was buried beneath continental and mountain glaciers up to one kilometre thick. However, by about 15,000 years ago, the ice sheets had begun their slow retreats to the west and northeast, dropping boulders and sediment as the margins of the glaciers melted. A narrow, ice-free corridor opened, running along the foothills. This windswept finger of exposed land and ice-dammed meltwater lakes determined how animals repopulated the interior of the continent as the Ice Age ended.

The ice sheets affected early Alberta's wildlife populations in other ways. So much water was locked up as ice during the glacial periods that sea level dropped. Large areas of shallow, offshore plain were exposed. An isthmus linking North America to Siberia opened. The Bering Land Bridge, a stretch of land about 1,500 kilometres wide, permitted animals to migrate between the continents. North American horses, mammoths, camels, and lions crossed to Asia, while animals such as wolves and elk moved to this continent. Going wherever there was food to eat, immigrating animals eventually spread throughout the Americas.

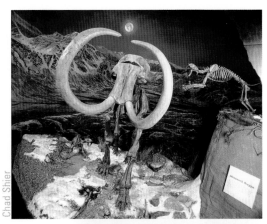

Chad Shier

Bruce Naylor, Royal Tyrrell Museum

The animals were pursued by a new kind of predator—*Homo sapiens*. The oldest known sites of human migrants from Asia are found in the Yukon and date from about 15,000 years ago. Research by the Geological Survey of Canada suggests that British Columbia's coastline may have been the route these early North Americans followed south. When Alberta's corridor opened between ice sheets, animals moved onto the interior plains of Canada, bringing human hunters in their wake.

Armed with stone, bone, and wooden tools, humans were efficient trackers and killers of prey. Even by 12,000 years ago, the time from which Alberta's oldest archaeological records of humans date, they may have altered their adopted land. Scientists are unable to say what caused the Great Extinction of the Quaternary Period, in which more than 40 kinds of large mammals disappeared from North America's plains. However, human hunters may have combined with rapidly changing climate and environments, and even with deadly new diseases imported by immigrating animals, to push ice-age mammals over the brink. During a span of 3,000 years, mammoths, sabretooth cats, dire wolves, North American horses, camels and llamas, bear-sized beavers, ground sloths, short-faced bears, and many other animals disappeared, ending the Age of Mammals.

As they left Alberta's Quaternary scene, other mammals took their places. The plains grizzly, plains wolf, and cougars became the region's large predators. Accounts by early Europeans travelling across Alberta during the eighteenth and nineteenth centuries tell of herds of bison stampeding across the plains like raging brown rivers, and of immense herds of pronghorn and caribou hunted by First Nations peoples.

These burrows were home to now-extinct prairie dogs that lived where the Hand Hills stand today. By living underground, prairie dogs used the snow as a blanket to keep their colonies warm: no matter how cold the temperature was in the open, under the snow it remained a constant 1° Celsius.

The creatures that lived in Alberta during the Ice Age were suited to cold climates. Size protected them: many were huge. Their increased mass and heavy coats of fur or hair helped to keep them warm. The woolly mammoth (top) used its great, curved tusks as plough blades, helping it clear pathways through deep snow and uncover grasses protected beneath.

Not all animals that lived during the Ice Age were huge, but even small creatures were big compared to modern relatives. Some beavers, fossils of which are found in the Yukon, were the size of bears. Prairie dogs were the size of house cats and lived in burrows such as these found in the Hand Hills area (bottom).

RECONSTRUCTING AN ICE-AGE ENVIRONMENT

Brian Kooyman

Water stored in the St. Mary's Reservoir north of Cardston insures area-farmers' crops against periods of drought, which are common in southern Alberta. In 1998, when the reservoir was drained to build a new spillway, researchers discovered tools used by ancient Albertans, as well as bones and footprints of at least 20 species of long-extinct animals.

"If you could have stood there 11,000 years ago, you could have seen mammoths, horses, camel, caribou, bison, wolves, foxes, ground squirrels, and birds—maybe all at once," says Len Hills, professor emeritus of geology and geophysics at the University of Calgary, and one of the scientists studying the site. Joining him in the research are university archaeologist Brian Kooyman and students Paul McNeil and Shayne Tolman, the site's discoverer. "We know from the way the trackways are intermingled that these animals were right there, together. And man was there, too—he was part of that environment."

The site is a window on the end of the Ice Age in southern Alberta. By inventorying the animals that lived in the region and studying the role of early humans at the site, Hills and colleagues reconstruct the ancient environment.

Wind was responsible for the burial and preservation of the St. Mary's Reservoir fossils 11,000 years ago, when it blew drifts of sand and dirt over the site from an ancient glacial-lake floodplain (top). Wind erosion exposed the site in 1998 after the reservoir had been drained. Wind is also the site's destroyer, stripping away sediment and trackways. At times, the scientists felt like they were racing the wind to preserve and collect the disappearing fossil material.

About 12,000 years ago, the site was an island delta at the end of a glacier-fed lake. When the St. Mary's River later drained the lake, it created a massive floodplain west of the island. Within a period of days or maybe weeks, large numbers of diverse animals visited the area to feed on the delta's lush vegetation and drink from the nearby river, trampling the soft ground with their hooves and feet. With so many animals congregating, humans were drawn to the area. Thick layers of sand and dirt blew eastward from the floodplain to bury tracks, bones, and tools. A series of brief geological moments were preserved, recording the presence of animals and humans over a 300-year period about 11,000 years ago.

The oldest tools are Clovis points, flaked-stone implements used by the first North Americans. Other tools date to the arrival of Europeans. Not far from where the team was excavating the skeleton of a horse, the researchers discovered tools that may have been used to kill the animal; laboratory tests reveal traces of ancient horse protein on two of the Clovis points.

"We've known from other sites in North America that they hunted mammoths," says Hills. "But this is the first solid evidence that Clovis people actually hunted horses. Maybe early humans influenced the extinction of these animals—not just mammoths, but horses, too."

What precisely caused the disappearance of so many Ice Age mammals 12,000 to 9,000 years ago is uncertain. Climates and regional environments were changing rapidly as the glaciers receded from the continent. Those changes alone would stress animal populations. New species from Asia may have increased competition for food, or introduced diseases. Humans may have been another factor in the ecological reorganization.

"The magnitude of the site," says Hills, "is the way that it is letting us see how these animals interacted with their environment, with each other, and with early humans."

Paul McNeil

Student researcher Paul McNeil inventoried and analyzed the trackways at the St. Mary's River fossil site. By comparing information about movement and size of the ancient animals with data from tracks made by modern elephant herds, he determined that the herds visiting the St. Mary's site 11,000 years ago were stressed; the ratio of young animals to adults was far below healthy levels.

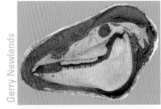

Gerry Newlands

Paul McNeil

Clovis hunting tools found in the same area as an almost complete fossil horse skeleton (top) retain residues of blood from *Equus conversidens*, an ancient horse that roamed southern Alberta at the end of the Ice Age. This is the first clear evidence that early North Americans hunted horses as well as mammoths.

The St. Mary's Reservoir fossil site contains remains and traces of at least 20 different kinds of Ice Age animals. Horses, camels, mammoths, musk oxen, bison, caribou, and other ungulates grazed in the region as the ice retreated to the northeast. By 9,000 years ago, most of these animals, including the camel, were extinct in North America (camel hoof print - bottom).

WHERE IN ALBERTA

Compared to the long history of the Earth, the last two million years are a geological heartbeat. Erosion has had little time to erase the footprints of the massive ice sheets from Alberta. Evidence exists throughout in the province—Ice Age calling cards left behind after the glaciers' visits.

Boulders that do not match nearby rock formations are scattered across Alberta. These out-of-place rocks, called glacial erratics, allow scientists to trace the path of long-disappeared ice sheets. Okotoks' Big Rock is made of quartzite from the Gog Group, a layer of rock that formed in the Jasper region during the Early Cambrian. It travelled to southern Alberta on the flank of Rocky Mountain glaciers during the Ice Age. Boulders of granite and gneiss found in fields near Viking arrived aboard the Laurentide ice sheet from the Northwest Territories.

Granite erratics high on the slopes of southern Alberta's Porcupine Hills indicate how thick the Laurentide ice sheet was when it covered that part of the province; 1,700 metres above prairie level. In the Sheep River valley, similar erratics are found at elevations of almost 1,400 metres.

Other glacial debris is seen along the Trans-Canada Highway between Cochrane and Seebe. Drumlins are mounds of rock and gravel shovelled and shaped by ice sheets as they pass over the land. The hills, which can be several kilometres long and hundreds of metres high, are teardrop-shaped. The blunt ends of the hills point in the direction that the ice was moving.

Other signs of retreating ice are easily seen in the Rocky Mountains. Moraine Lake, for instance, is dammed by gravel ridges that give the lake its name. High overhead, the jagged peaks of the surrounding mountains attest to the ability of ice to carve rock. Before the Ice Age, the Rocky Mountains featured fewer crags and peaks, and fewer wide, U-shaped valleys.

Dennis Braman, Royal Tyrrell Museum

Country

Parks & Protected Areas, Kananaskis

Wil Andruschak

Alberta is littered with evidence of the Ice Age: giant boulders carried on the ice and left far from their sources, such as the Big Rock at Okotoks (top); deposits of glacial gravels and till in the form of drumlins (middle); and moraines (bottom) marking edges of the former glaciers.

WHERE IN ALBERTA

Potholed landscapes north of Drumheller and in Elk Island National Park tell of icebergs calving off massive glaciers into huge, dammed glacial lakes as the Ice Age ended. When the lakes drained, some of frozen chunks stayed behind and were blanketed in sediment. Eventually, the ice cubes melted and the overlying sediment collapsed, forming depressions called kettles. Today, Alberta's knob-and-kettle landscapes are characterized by ponds and sloughs surrounded by aspen parkland and pastures. The potholes are important stopovers for migrating waterfowl.

The kind of vegetation that grows in an area reflects the kind of rock that lies beneath. For a view of the ice sheets' effect on southern Alberta's soils, visit Cypress Hills Provincial Park, one of the few areas in the province that was not submerged by ice. The lodgepole pine forest is rooted in early and mid-Tertiary gravels. White spruce grows where older Cretaceous-aged rock lies at the surface. Surrounding the hill-islands, the province's grasslands are matted atop glacial till—dirt, gravel, and rock picked up by ice sheets, mixed, and carried great distances before being spread across the prairies as if it were stony fertilizer. In the Drumheller Valley, the till is the soft, yellowish, topmost layer in the valley walls, ranging from 10 to 20 metres thick.

The courses of the Bow, Red Deer, Old Man, and Milk rivers chart how each river was rerouted southward across the plains by retreating ice sheets. In Edmonton and Calgary, the terraced slopes of the Bow, Elbow, and North Saskatchewan river valleys trace the glacier's slow withdrawal in those regions. The Red Deer River's wide, sculpted valley through Drumheller and Dinosaur Provincial Park tells of the sudden draining of Glacial Lakes Drumheller and Bassano. The sediments blown from the bottom of another drained glacial lake, this one north of Cardston, helped preserve the ice-age fossils, trackways, and tools being studied by Len Hills and colleagues at the Milk River Reservoir.

Tim Schowalter

Royal Tyrrell Museum

Dennis Braman, Royal Tyrrell Museum

The prairie ponds that dot the knob-and-kettle landscape (top) north of Drumheller are a result of giant icebergs breaking off the edges of retreating ice sheets, being buried in sediment, and then melting.

Glacial Lakes Drumheller and Bassano carved out different parts of the Red Deer River Valley (middle) during the few days it would have taken the lakes to drain.

The Cypress Hills (bottom) demonstrate the effect of soil type on vegetation in an area. Forests cover the older rocks of the hills, while below, glacial tills support prairie grasslands.

ALBERTA TODAY

Courtesy of Economic Development Edmonton

Albertans now live in the Holocene Epoch, the most recent chapter in the Quaternary Period. It is a chapter still being written. During this fragment of Earth's history, human time intersects with the long, slow pace of geological time, and Albertans' dependence on what has gone before becomes clear. Most of the province's wealth and industry is built upon the past: coal mining on Jurassic swamps and bogs, petroleum extraction on Devonian-aged ocean reefs, tourism upon Cretaceous dinosaurs and ice-sculpted mountains, agriculture and forestry on soils deposited during the last ice age. Modern Alberta is merely a collection of environments that existed here long ago, which have been reworked and preserved within the geological record.

Modern Albertans are adding to the record. During the last 150 years, the rate of change by human numbers and behaviour on Alberta's landscape has increased. The presence of more than two million people living in the province has changed the land and affected its nonhuman residents. Some Alberta plants and animals have joined the mass-extinction occurring planet-wide: gone are the plains grizzly and the plains wolf, and bison exist only in small numbers in captivity and in parks such as Alberta's Wood Buffalo National Park and Elk Island National Park. Our highways, mines, and cities leave their own traces—recording the presence of Albertans in this place at this time as trackways in a Grande Cache coal mine chronicle the long-ago passage of dinosaurs, or as burrows in rocks along the front ranges of the Rocky Mountains mark ancient homes of Triassic worms and clams.

As plants and animals that lived in the province millions of years ago did, today's Albertans are creating their own legacy of artifacts and traces which will be preserved through time.

RECOMMENDED READINGS

Geology & Palaeontology

Callaway, Jack M., and Elizabeth L. Nicholls, eds. *Ancient Marine Reptiles.* Academic Press, 1997.

Cattermole, Peter. *Building Planet Earth.* Cambridge University Press, 2000.

Coppold, Murray, and Wayne Powell. *A Geoscience Guide to the Burgess Shale: Geology and Palaeontology in Yoho National Park.* Yoho-Burgess Shale Foundation, 2000.

Cowen, Richard. *The History of Life*, 3rd ed. Blackwell Science Press, 2000.

Gould, Stephen J. *Wonderful Life: The Burgess Shale and the Nature of History.* W.W. Norton & Co., 1989.

Ludvigsen, Rolf, ed. *Life in Stone.*

Pielou, E.C. *After the Ice Age: The Return of Life to Glaciated North America.* University of Chicago Press, 1991.

Schopf,. J. William. *Cradle of Life.*

Dinosaurs

Currie, Philip J., and Kevin Padian, eds. *Encyclopedia of Dinosaurs.* Academic Press, 1997.

Currie, P.J. and J. Sovak. *The Flying Dinosaurs.* Red Deer Press, 1986.

Farlow, James O., and M.K. Brett-Surman, eds. *The Complete Dinosaur.* Indiana University Press, 1997.

Keiran, Monique, and the Royal Tyrrell Museum. *Albertosaurus: Death of a Predator.* Raincoast Books, 1999.

Keiran, Monique, and the Royal Tyrrell Museum. *Ornithomimus: Pursuing the Bird-Mimic Dinosaur.* Raincoast Books, 2002.

Lockley, Martin. *Tracking Dinosaurs: A New Look at an Ancient World.* Cambridge University Press, 1991.

Russell, D.A. *An Odyssey in Time: The Dinosaurs of North America.* University of Toronto Press, 1989.

History of Alberta Palaeontology

Spalding, David E. *Into the Graveyard of the Dinosaurs: Canadian Digs and Discoveries.* Doubleday Canada, 1999.

Sternberg, C.H. *Hunting Dinosaurs in the Badlands of the Red Deer River, Alberta, Canada.* NeWest Press, 1985.

Exploring Alberta's Ancient Past

Gadd, Ben. *Handbook of the Canadian Rockies*, 2nd ed. Corax Press, 1995.

Gross, Renie. *Dinosaur Country: Unearthing the Alberta Badlands.* Badlands Books, 1998.

Huck, Barbara, and Doug Whiteway. *In Search of Ancient Alberta: Seeking the Spirit of the Land.* Heartland Publishers, Winnipeg, 1998.

Mussieux, Ron, and Marilyn Nelson. *The Traveller's Guide to the Geological Wonders of Alberta.* Federation of Alberta Naturalists and Canadian Society of Petroleum Geologists, 1998.